Allee Effects in Ecology and Conservation

Allee Effects in Ecology and Conservation

Franck Courchamp
Laboratoire Ecologie Systematique et Evolution,
CNRS University Paris Sud, Orsay, France

Luděk Berec
Department of Theoretical Ecology, Institute of Entomology,
Biology Centre ASCR, České Budějovice, Czech Republic

Joanna Gascoigne
School of Ocean Sciences, University of Wales Bangor,
Menai Bridge, UK

OXFORD
UNIVERSITY PRESS

OXFORD
UNIVERSITY PRESS

Great Clarendon Street, Oxford OX2 6DP

Oxford University Press is a department of the University of Oxford.
It furthers the University's objective of excellence in research, scholarship,
and education by publishing worldwide in

Oxford New York

Auckland Cape Town Dar es Salaam Hong Kong Karachi
Kuala Lumpur Madrid Melbourne Mexico City Nairobi
New Delhi Shanghai Taipei Toronto

With offices in

Argentina Austria Brazil Chile Czech Republic France Greece
Guatemala Hungary Italy Japan Poland Portugal Singapore
South Korea Switzerland Thailand Turkey Ukraine Vietnam

Oxford is a registered trade mark of Oxford University Press
in the UK and in certain other countries

Published in the United States
by Oxford University Press Inc., New York

British Library Cataloguing in Publication Data

Data available

Library of Congress Cataloging in Publication Data

Data available

Typeset by Newgen Imaging Systems (P) Ltd, Chennai, India
Printed in Great Britain
on acid-free paper by
Biddles, Ltd, King's Lynn

ISBN 978–0–19–857030–1

10 9 8 7 6 5 4 3 2 1

Foreword

When I heard of this party, before committing myself to saying yes or no, I asked Kate whether she would go. She said she had already asked Mike and Pedro, and neither of them were going, mostly because Aisha couldn't make it; so she wouldn't either. She would go with her buddies Petra and Lynn instead, to watch a movie or something. Then I went to see Andrea, because at least he always has a bunch of people hanging around, and would hopefully bring them along. But Andrea was out doing something else, and wouldn't go to that one, because he heard nobody was going anyway. Counting how few were going, I realized it was well below the threshold to have a blasting wild party, and to attract more people. So of course, I didn't go either. I went to the mall instead, to rent a DVD. Thus, I didn't eat and drink all that free party stuff, and I came back home alone. Bad for my fitness. All this because there were not enough people to start with... But that's alright, I'll go to Amanda's next week, I heard everybody will be there...

Preface

The more the merrier; everybody has heard that maxim. This book is about its equivalent in ecology: the Allee effect. This effect is simply a causal positive relationship between the number of individuals in a population and their fitness. The more individuals there are (up to a point), the better they fare. One may think that this simple concept is a minor detail in theoretical population dynamics, but in fact it has a surprising number of ramifications in different branches of ecology: population, and community dynamics, behavioural ecology, biodiversity conservation, epidemiology, evolution, and population management.

The Allee effect is an ecological concept with roots that go back at least to the 1920s, and fifty years have elapsed since the last edition of a book by W.C. Allee, the "father" of this process. Throughout this period, hardly a single mention of this process could be found in ecological textbooks. The concept lurked on the margin of ecological theory, overshadowed by the idea of negative density dependence and competition. The situation has appeared to change dramatically in the last decade or so, however, and we now find an ever-increasing number of studies from an ever-increasing range of disciplines devoted to or at least considering the Allee effect. It was only natural that this boom would sooner or later require a monograph that presents this concept and its implications, and we hope that we have produced a fair synthesis.

The book aims to take an overarching view of all of the branches of ecology which nowadays embrace the Allee effect, including the empirical search for Allee effects, theoretical implications for long-term population dynamics, its role in genetics and evolution and its consequences for conservation and management of plant and animal populations. This diversity means that the book is (we hope) of interest to empiricists, theoreticians and applied ecologists, as well as to students of ecology and biodiversity and to conservation managers. Such a diverse potential audience inevitably means that we cannot suit everyone perfectly as to form and content. Any of these three aspects of Allee effects (the empirical, the theoretical, the applied) could probably fill a volume the same length as this book, so tradeoffs were necessary to avoid scaring readers with a monstrously huge volume. We chose to address all relevant aspects of Allee effects while trying to maintain a balance between conciseness and comprehensiveness. We recognize that our diverse readers will have diverse interests and needs, so we have chosen to present distinct chapters on each topic, confining, for example, all mathematical equations to one chapter and all genetic and evolutionary discussions to another. Our goals have been to present the Allee effect concept to the reader with clarity, to demonstrate its ubiquity, its diversity and its importance in plant and animal species, and finally to help researchers in specific branches of ecology to look into each others' kitchens.

Another aim has been to produce a book that requires little background knowledge, aside from curiosity and some basic notions of ecology. Of course, the book is not an elementary text in ecology, genetics, evolution, or mathematical modelling of population dynamics, so that the reader who handles the basics of these disciplines will profit most from the book. But we have tried to explain concepts in clear language, and where ecological jargon becomes necessary, we provide a glossary to carry the reader through the relevant chapter. We hope that this first book on the Allee effect for fifty years achieves these aims, and mostly that you enjoy it.

Acknowledgements

Many people helped polish previous book drafts, or draft chapters, by providing invaluable comments or criticism. We would like to thank, in no particular order, Sergei Petrovskii, Tobias van Kooten, Philip Stephens, David Boukal, Vlastimil Krivan, Frantisek Marec, Martina Zurovcova, Mike Fowler, John Cigliano, Julia Jones, Jan Hiddink, Michel Kaiser, Nick Dulvy, Alain Morand, Johan Chevalier, Xavier Fauvergue, Eric Vidal, Carmen Bessa-Gomez, Emmanuelle Baudry, Vincent Devictor, Gwenaël Jacob, Elena Angulo and Xim Cerda, with profuse apologies to anyone we have forgotten. All errors and omissions remain the responsibility of the authors. We thank John Turner and Greg Rasmussen for providing photographs. Luděk Berec acknowledges funding from the Institute of Entomology of the Biology Centre ASCR (Z50070508) and the Grant Agency of the Academy of Sciences of the Czech Republic (KJB600070602). Franck Courchamp acknowledges funding from the Agence Nationale de la Recherche. Jo Gascoigne acknowledges funding from the UK Biotechnology and Biological Sciences Research Council.

Contents

1. *What are Allee effects?*

1.1. Competition versus cooperation

It is obviously difficult to pinpoint what will be the major contribution of a researcher centuries after his death, but after only a few decades one thing is already sure: Professor Warder Clyde Allee will remain among those who will have clearly marked ecological sciences. And not only once. Indeed, ironically, it is not the Allee effect which made him most well known. Although he has always been quite famous among ecologists, Allee has for most of the last seventy years enjoyed greater renown among behavioural ecologists than population dynamists or conservation biologists. His work has been among the most influential for animal behavioural research. In fact, one of the highlights of the annual *Animal Behaviour Society* meetings, a scientific competition where a dozen of the most promising young researchers in the field present their work to an audience of established specialists, is called the Allee Competition.

Allee was born on 5 June 1885 near Annapolis, Indiana, in a small Quaker community. He began and finished his career in Chicago, forming the basis of what would become 'the Chicago School of Behaviour' (Banks 1985). Allee was among the first to study species from an ecological perspective and had a major role in the establishment of ecology as an independent biological discipline. He was also one of the first ecologists to use statistics.

Allee's main interest was to ascertain the factors leading to the formation and continuation of animal aggregations. His idea was that such aggregations, which increased individual survival, were an initial stage in the evolution of sociality. Because he thought the animals were not conscious of such benefits, he often preferred the term 'proto-cooperation' to the more anthropomorphic 'cooperation'.

Allee's hypotheses of favourable selection of cooperation hinged mostly upon observations that certain aquatic species can affect the chemistry of the water, by releasing protective chemicals such as calcium salts that could enhance their survival. Evidently, this process was effective when the chemicals were not too diluted, that is, when the animals were aggregated. His experiments showed that goldfish were better able to render the water closer to their optimal chemical requirements when several were in the tank (Allee *et al.* 1932), and that planarian worms and starfish could better survive the exposure to UV or toxic chemicals when they were numerous enough to condition the water (Allee *et al.* 1938, 1949b).

Rapidly enough, Allee gathered sufficient experimental and observational data to conclude that the evolution of social structures was not only driven by competition, but that cooperation was another, if not the most, fundamental principle in animal species (Allee 1931). The dynamical consequences of this importance of animal aggregations directly led to what Odum called in 1953 'the Allee principle', now known as the Allee effect (Odum 1953).

Allee was not shy in applying his theories to humans, believing in particular that in them, as in many species, individual survival was enhanced by a propensity for cooperative tendencies which outweighed competition and aggression. As was later pointed out to him, his upbringing in a peace-loving, Quaker environment may have deeply influenced Allee's theories. He refuted this and it is difficult to know whether or not it is true. Yet, there will always remain a doubt that had not Allee deeply believed in the importance of cooperation in mankind, you would not be reading this book on Allee effects now.

Figure 1.1. Warder Clyde Allee in his University of Chicago office, reading our book with great interest © Special Collections Research Center, University of Chicago Library.

From the behaviour of animal groups to its dynamical consequences, the path is short. It may thus appear surprising that the development of behavioural studies on cooperation versus competition has not been mirrored in population dynamics, which has developed a focus almost exclusively on the principles of competition. Up to now. Indeed, it seems that Allee's work is once again starting to have a major influence on fundamental and applied ecology, this time from the population dynamics perspective. As happened in behavioural studies, the notion that cooperation, in its broad sense, can be as important as competition is changing our views of population dynamics and all related ecological fields.

1.2. The influence of density in population dynamics

When a population is small, or at low density, the classical view of population dynamics is that the major ecological force at work is the release from the

Figure 1.2. W. C. Allee used goldfish in tanks to demonstrate survival rates in groups © Michael Somers.

constraints of intraspecific competition. The fewer we are, the more we all have, and the better each will fare. As each of the few individuals benefits from a greater availability of resources, the population will soon bounce back to intermediate densities. This view, although often the only one taught to generations of biologists, is however lacking a crucial component: cooperation.

The individuals of many species cooperate, *sensus largo*: they use cooperative strategies to hunt or to fool predators, they forage together, they join forces to survive unfavourable abiotic conditions, or simply they seek sexual reproduction at the same moment and/or place. When there are too few of them, it may be that they will each benefit from more resources, but in many cases, they will also suffer from a lack of conspecifics. If this is stronger than the benefits, then the individuals may be less likely to reproduce or survive at low population size. Their fitness may then be reduced. In these instances, fitness and population size or density are related: the lower the population size or density, the lower the fitness. This, in essence, is an Allee effect.

Formal definitions of Allee effects encompass this notion of positive density dependence: the overall individual fitness, or one of its components, is positively related to population size or density. Most definitions of the Allee effect apply either to the density of the population or to its size. For the sake of clarity, we will speak of density, meaning indifferently population size or density. There are cases where size and density are not interchangeable, however, and in these cases, we will be more specific. Chapter 2 will provide examples of mechanisms operating only on population size, only on population density, or on both. We stress here that the word 'or' in 'population size or density' is very important: even large populations can be exposed to Allee effects when sparse, or dense populations when small.

In a few paragraphs, we will move on to more formal definitions. But the basic concept of the Allee effect does not require complex explanation of subtle details, and is very easy to understand when put in simple words. If, for example, adults reproduce less well when less numerous, then each of these fewer adults will make fewer offspring. And these fewer offspring will grow to become fewer adults. This in turn will make even fewer offspring at the next generation, until the population goes extinct. The same works with survival: imagine a species that fares better when there are a minimum number of individuals, say, to avoid predation or to acquire resources. Then, when there are too few individuals, these too few will survive less well. Their population will drop even further, and expose them to even lower survival, until the population goes extinct. These are extreme cases of a mechanism that can be more subtle and especially more masked. Yet all Allee effect concepts are variants of these simple examples.

1.3. **Studies on the Allee effect**

When the first examples of Allee effects were proposed by WC Allee, they were not any more complex than the two proposed above. Even at this early stage they rapidly covered a large span of the animal kingdom (no plant examples were proposed yet; Allee was a zoologist), but they were often limited to description without any proposed explanatory mechanism. For example, in the struggle against tsetse fly (*Glossina sp.*), it was observed that below a minimum density, flies were reported to disappear spontaneously from the area, but no mechanism was proposed. Early laboratory experiments showed that higher densities of fertilized eggs of sea urchins or frogs led to their accelerated development (Allee *et al.* 1949b) or that grouped rotifers (*Philodina roseola*) survived chemical toxin better than single individuals (Allee *et al.* 1949a), but again the mechanism was not understood.

Among the first mechanisms proposed to explain higher survival in larger groups was the protection provided by animal aggregations from external threats. In many cases, aggregations reduce the total surface (relative to the total mass) exposed to chemicals, extreme weather, bacteria or predators. Examples include bobwhite quails (*Colinus virginianus*) who huddle together to lower the surface presented to cold weather and American pronghorn (*Antilocapra americana*) forming a defensive band that presents a minimum group surface to wolves and coyotes (Allee *et al.* 1949b).

Figure 1.3. Bobwhite quails huddling together to fight low temperatures during winter. The smaller the group, the larger the surface of each member exposed to the outer environment, and the higher the mortality during winter © Stuhr Museum of the Prairie Pioneer.

Shortly after, analysing data on the flour beetle, *Tribolium confusum*, Allee observed that the highest per capita growth rates of their populations were at intermediate densities (Fig. 1.4). That they were lower at high density was not surprising: intraspecific competition was too high, in this case taking the form of cannibalism of eggs by adults. A less expected finding was that fewer matings led to insufficient stimulation of females. As a result, when fewer mates were present, the females produced fewer eggs, which is not an obvious correlation for an insect. In this case, optimal egg production was thus achieved at intermediate densities.

The initial studies on Allee effects were thus experimental, and done by Allee himself and by his students. The concept was not much embraced by the wider ecological community, and for half a century only a couple of studies a year were related to the subject, with the mainstream of research focusing on classical negative density dependence. Among these early studies, most were theoretical, probably for two reasons. First, demonstrating an Allee effect in the wild usually requires a long and complete data set of population dynamics; logistics generally precluded the collection of this type of data at the time and models were thus less challenging to produce. Second, population dynamicists were not interested in undercrowding, which was regarded as an interesting but anecdotal process. The history of research into Allee effects has been concisely but nicely summarized in Dennis (1989).

As a result, there were fewer than 50 studies mentioning the Allee effect *per se* at the turn of the millennium. It was not until the biodiversity crisis started to

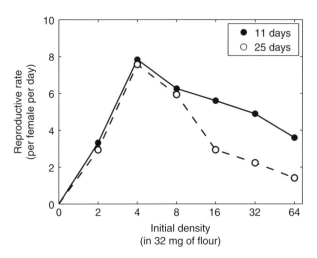

Figure 1.4.   Relationship between initial population density of the flour beetle, *Tribolium confusum*, (number of adults in 32 mg of flour) and per capita population growth rate (per female per day) during 11 (full line) or 25 days (straight line). Modified from Allee *et al.* (1949b). Copyright: Clemson University — USDA Cooperative Extension Slide Series, Bugwood.org

draw the attention of fundamental ecologists, and after much delay, that this concept was revived, rapidly increasing this number by an order of magnitude (Fig. 1.6). With so many small and declining populations, it became important to focus on to why populations get smaller or sparser in the first instance (the 'declining population paradigm'; Caughley, 1994) and what happens when individuals get fewer or are driven to lower densities (the 'small population paradigm'; Caughley, 1994). So, in the late nineties, the concept started to grab the interest of researchers. Cooperation was making its way into population dynamics, and consequently, in related applied domains such as conservation biology, biological control and fisheries management.

The studies of the early nineties, before the real start of the upsurge, mostly concerned marine invertebrates (Levitan *et al.* 1992, 1995, Pfister *et al.* 1996), plants (Hoddle 1991, Lamont *et al.* 1993, Widén 1993) and insects (Hopper *et al.* 1993, Fauvergue *et al.* 1995), with only very few studies on vertebrates, with the exception of the fisheries literature (Lande *et al.* 1994, Myers *et al.* 1995, Saether *et al.* 1996) but see also Veit *et al.* (1996). During this time, a few theoretical studies were also published (e.g. Dennis 1989, Knowlton 1992, Lewis *et al.* 1993, McCarthy 1997), some of them quite influential. The revival of the interest in Allee effects was noticed at the end of the century

Figure 1.5. Confused flour beetles, *Tribolium confusum*, are some of the most abundant and injurious insect pests to grain products and among the first to demonstrate an Allee effect.

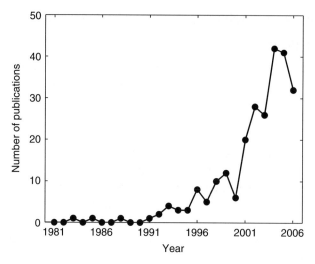

Figure 1.6.   Number of publications in peer-reviewed journals mentioning 'Allee effect' in title, keywords or abstract in the last 25 years (source: ISI Web of Knowledge).

and emphasized in two reviews on Allee effects published back to back in the same journal in 1999 (Courchamp *et al.* 1999a, Stephens and Sutherland 1999). From then, the Allee effect quickly gained recognition and interest. So, in a sense, studies on the Allee effect have shown a pattern quite similar to... an Allee effect. After a long lag of barely noticed presence, not unlike some invading organisms that will be discussed later in the book, studies on Allee effect started to increase at an accelerating rate. The conceptual resemblance between the lag-time of Allee effect studies after the concept was set out and the lag-time of invasions of species subject to Allee effects after a founder population arrives is indeed amusing. We might speculate that most new fundamental concepts may be essentially subject to such a process: there is a critical number of studies required for the topic to flourish, so that it needs some time to gain interest and only then starts to increase in numbers of papers, exponentially or even quicker.

   After the dominance of experimental and theoretical studies, the studies are now rushing in two different directions: theoretical, with mathematical models of ever increasing details; and empirical, with the unveiling of Allee effects in natural populations. As we will see, the current state of the research is sufficiently documented to demonstrate unambiguously the ubiquity of Allee effects in an impressive range of taxa and ecosystems. Yet, it still may not be sufficiently documented to provide a true representation of its importance.

1.4. **What is and what is not an Allee effect**

1.4.1. **Definitions**

With the development of this procession of studies, inevitably comes a whole suite of nuances and subtleties, as we progress in our understanding of Allee effects and are able to draw fine distinctions. There are many concepts related to Allee effects, and the definitions of each of them ought to be precise and clear so as not to create undue confusion between them. Below, we present some definitions, with the aim of providing a coherent and homogenous textbook, but at the same time with the knowledge that these are our own acceptance of these terms, and that other definitions may be found that are more suited. Allee did not give any specific instructions, nor did Odum when he introduced the term 'Allee principle'. This is also why we need to provide some clear definitions, and refine them with time, as research advances and new cases occur, so that the whole community eventually agrees as to what is or is not an Allee effect. We thus provide here a proposition of classifications and categorizations. For the sake of precision, we present them in the form of simple definitions (Box 1.1), but they are also explained in the text.

Other terms have been used to describe the Allee effect. As the classical relationship between individual fitness and population density is negative (fitness decreases with increasing density, Fig. 1.7a), the most correct technical term for Allee effects is 'positive density dependence' (individual fitness is positively correlated with density, see Fig. 1.7b). It is also sometimes confusingly called 'inverse density dependence' (it has the inverse relationship to density compared to classical dynamics; we feel this term is confusing and best avoided). The fisheries literature uses depensation and depensatory dynamics, probably by analogy with 'compensation' and 'compensatory dynamics'. Some scientists also refer to the underpopulation effect or undercrowding, as the opposite of processes occurring through overpopulation or overcrowding.

1.4.2. **Component and demographic Allee effects**

Within the Allee effect domain, the most important distinction is between component and demographic Allee effects (Stephens *et al.* 1999). The distinction is very simple. Component Allee effects are at the level of components of individual fitness, for example juvenile survival or litter size. Conversely, demographic Allee effects are at the level of the overall mean individual fitness, practically always viewed through the demography of the whole population as the *per capita* population growth rate. The two are related in that component Allee effects may result in demographic Allee effects. Classically, individual fitness is assumed higher at low densities because of lower intraspecific competition (Fig. 1.8a). When the individuals have a lower value for one fitness component at low density (e.g. adult survival, Fig. 1.8b,c), the advantages of lower competition

Box 1.1.   **Definition of the main Allee effect-related concepts of the book (adapted from Berec *et al.* 2007). Although not all the concepts in this box are mentioned in Chapter 1, this box aims to serve as a reference point and the reader may go back to it when appropriate.**

Allee threshold: a critical population size or density below which the per capita population growth rate becomes negative (see Fig. 1.9).

Component Allee effect: a positive relationship between any measurable component of individual fitness and population size or density.

Demographic Allee effect: a positive relationship between the overall individual fitness, usually quantified by the per capita population growth rate, and population size or density.

Depensation: the term often used in place of Allee effects in the fisheries literature.

Dormant Allee effect: a component Allee effect that either does not result in a demographic Allee effect or results in a weak Allee effect and which, if interacting with a strong Allee effect, causes the overall Allee threshold to be higher than the Allee threshold due to the strong Allee effect alone.

Emergent Allee effect: In population models, demographic Allee effects that emerge without treating an underlying component Allee effect explicitly, i.e. without using any component Allee effect model.

Multiple Allee effects: any situation in which two or more component Allee effects work simultaneously in the same population.

Nonadditive Allee effects: multiple Allee effects that give rise to a demographic Allee effect with an Allee threshold greater or smaller than the algebraic sum of Allee thresholds owing to single Allee effects.

Sigmoid dose-dependent response: many (asexually reproducing) macroparasites and many microparasites need to exceed a threshold in load to inflict any harm on their hosts or to spread effectively in the host population; this behaviour may be considered a kind of demographic Allee effect in parasites.

Subadditive Allee effects: multiple Allee effects that give rise to a demographic Allee effect with an Allee threshold smaller than the algebraic sum of Allee thresholds owing to single Allee effects.

Box 1.1. *(Continued)*

Superadditive Allee effects: multiple Allee effects that give rise to a demographic Allee effect with an Allee threshold greater than the algebraic sum of Allee thresholds owing to single Allee effects.

Strong Allee effect: a demographic Allee effect with an Allee threshold (see Fig. 1.9).

Weak Allee effect: a demographic Allee effect without an Allee threshold.

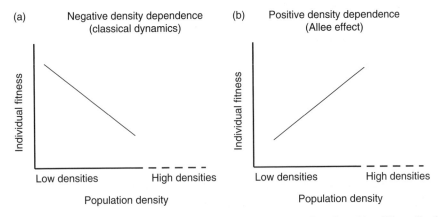

Figure 1.7. A schematic sketch of negative (classical dynamics) and positive (Allee effect) relationships between individual fitness and population density.

often compensate through a higher value of one or more other fitness components (e.g. juvenile survival, Fig. 1.8b). The end result is classical negative density dependence (Fig 1.8d). Alternatively, if compensation is insufficient, the effect of lower individual fitness values at low density prevails and will thus result in a lower *per capita population* growth rate: there will be a demographic Allee effect (Fig. 1.8e).

A demographic Allee effect is a proof of the existence of at least one component Allee effect, although not one which is necessarily obvious. The tsetse flies probably had a demographic Allee effect, but the component Allee effect(s) had not been identified. By contrast, a component Allee effect in a population may not always generate a demographic Allee effect. For example, the island fox (*Urocyon littoralis*) is a territorial, monogamous species. Upon the death of a mate, the survivor seeks a new mate among the individuals of the neighbouring territories. If

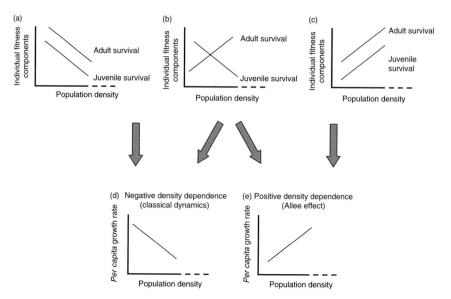

Figure 1.8. Illustration of the correspondence between interacting components of individual fitness and the resulting types of density dependence. While (a) always gives (d) and (c) always gives (e), (b) can give either (d) or (e).

density is too low, the probability of finding a mate declines: a component Allee effect (Angulo *et al.* 2007). However, this Allee effect is not sufficient to counteract the release from intraspecific competition at low density. When density is low, fewer female foxes reproduce, but those who do have more resources and will produce larger litters (Angulo *et al.* 2007). As a result, the per capita reproduction is negatively density dependent. Unless there is another, intense component Allee effect in survival, there will be no demographic Allee effect. The *Tribolium* species mentioned above showed a component Allee effect leading to a demographic Allee effect, while the goldfish only showed component Allee effects. The Caribbean long-spined urchin, *Diadema antillarum,* has a component Allee effect due to a lower fertilization efficiency at low density (see Chapter 2), but this is compensated by an increase in the gamete output under reduced competition for food, resulting in absence of demographic Allee effect (Levitan 1991).

The result of a demographic Allee effect is a downward bending of the relationship between population density and the *per capita* population growth rate at low density, so that, accounting for interspecific competition at high density, the relationship is hump-shaped (Fig. 1.9). From this, one can further distinguish two different kinds of demographic Allee effects: the weak and the strong (Wang *et al.* 2001).

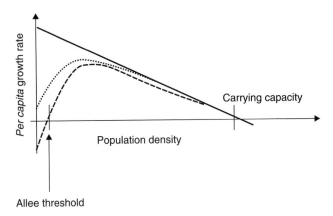

Figure 1.9. Classical negative density dependence (solid) compared to strong (dashed) and weak (dotted) Allee effects. '*Many have written pointedly about over-crowding, and while there is still much to be learned in that field, it is in the recently demonstrated existence of under-crowding, its mechanisms and its implications that freshness lies. Without for one minute forgetting or minimizing the importance of the right-hand limb of the last curve, it is for the more romantic left-hand slope that I ask your attention.*' (Allee 1941).

1.4.3. **Weak and strong Allee effects**

A demographic Allee effect is weak if at low density the per capita population growth rate is lower than at higher densities, but remains positive (dotted curve on Fig. 1.9). On the contrary, it can become so low as to become negative below a certain value, called the Allee threshold (dashed curve on Fig. 1.9): it is then a strong Allee effect. If a population subject to a strong Allee effect drops below that threshold, the population growth rate becomes negative and the population will get smaller at an accelerating rate (due to a continuing decrease in the per capita population growth rate), until it reaches zero, or extinction.

Empirically, it is easier to demonstrate that an Allee effect is strong (i.e. that a threshold exists) than that it is weak (i.e. that there is no threshold, and not that the study fails to reveal one). With adequate time series, one can pinpoint the density for which the per capita population growth rate becomes negative. Demonstrating a threshold, however, is no simple task. First, estimates of the Allee threshold are usually approximate because low abundance will yield high observation error, fluctuations due to demographic stochasticity and a significant proportion of counts at zero (Johnson *et al.* 2006). Despite an impressive dataset, the study on the island fox discussed above could not identify the value of the Allee threshold even though the Allee effect was strong in some populations because these populations were already declining (Angulo *et al.* 2007). A strong Allee effect has also been shown in the gypsy moth (*Lymantria dispar*), an invasive pest spreading across eastern North America.

With the study of the spatiotemporal variability in rates of spread of this species, Johnson *et al.* (2006) were able to estimate an Allee threshold. In contrast, another invasive species, the smooth cordgrass, *Spartina alterniflora,* was shown to display a weak Allee effect, with a reduced, yet still positive, per capita population growth rate in sparsely populated areas, because even single individuals were able to grow vegetatively or self-fertilize (Davis *et al.* 2004, Taylor *et al.* 2004).

1.4.4. **On the margins of the Allee effect**

Until a clear definition of what constitutes an Allee effect was presented (Stephens *et al.* 1999), a number of mechanisms were suggested which some researchers considered mechanisms of an Allee effect while the others did not. Unfortunately, some controversy, or misunderstanding, still persists. This is not to say that an alternative definition of what is and what is not an Allee effect cannot be proposed, which may then modify our perception of what constitutes an Allee effect mechanism. To stay on firm grounds, we recall that we define Allee effects as a *causal* positive relationship between (a component of) *individual* fitness and either population size or density.

A number of controversies concern the word 'individual'. An example comes from genetics. Could genetic drift and inbreeding be considered to generate Allee effects? An early response, embraced by many authors, is that only processes that are entirely linked to demographic mechanisms should be regarded as Allee effects. Yet, as we discuss later in this book, genetic factors can contribute to reduced individual fitness in small or sparse populations, and as such they fall under the definition of Allee effects we follow.

Another controversial issue concerns demographic stochasticity. Is it an Allee effect mechanism or not? Demographic stochasticity refers to the variability in population growth rates arising from the random variation among individuals in survival and reproduction (and is often restricted to the variation in numbers or gender of the offspring). Obviously, demographic stochasticity increases vulnerability of small populations to extinction, but the key point is that it does not reduce mean *individual* fitness as a population declines. Demographic stochasticity in birth and death rates thus do not create an Allee effect. Neglecting for the moment other density-dependent mechanisms, this is because the intrinsic probability of an individual reproducing or surviving a given period is not in this case a function of population size. The exception to this, in our opinion, is stochastic fluctuations in the (adult) sex ratio, which is also considered a type of demographic stochasticity and which does indeed lead to a reduction in individual fitness as population declines. Unlike other forms of demographic stochasticity, this can thus be considered an Allee effect mechanism (Stephens *et al.* 1999,

Møller and Legendre 2001, Engen *et al.* 2003). We discuss this relatively thorny point further in the final book chapter.

Another controversy, or rather misunderstanding, concerns the word 'causal', as in 'a *causal* positive relationship' between fitness and density. Here we can clearly point out cases where would-be Allee effects might be misleading. It is important to realize that a positive relationship between density and fitness cannot thoughtlessly be claimed an Allee effect, because other factors can be present which control both density and fitness. An obvious example is that low population density and low individual fitness might often be positively correlated via poor habitat quality. This looks superficially like an Allee effect (individuals in sparse populations suffer reduced fitness) but in fact is not, because there is no causal link between fitness and density. The test is to ask this: if the high density population were reduced to low density, would fitness of its individuals decrease? If the answer is 'yes', an Allee effect is present. In this example, however, the answer would be 'no', because the population would still be in high quality habitat. This problem of confounding variables occasionally arises in studies which purport to show Allee effects, and one should always be on guard against it.

The opposite may also be true: Allee effects may act where unsuspected. It is important to keep in mind that for some species Allee effects do not only occur in small or low-density populations. The behaviours and ensuing dynamics that generate an Allee effect when there are few conspecifics are generally intrinsic at the species level and shape its life history traits all along the spectrum of population sizes or densities. For some species, Allee effects can have consequences at any size or density. A population may thus be exposed to a decline even before we realize that it is actually in trouble.

1.5. **Allee effects in six chapters**

All these concepts and rhetorical definitions may lead to a false impression of distance from the reality of applied ecology and of remote importance for concrete cases of conservation biology. On the contrary, Allee effects are very tangible mechanisms, likely to affect persistence of populations in very substantial ways. It is of the utmost importance to realize that many taxa in many ecosystems are sensitive to Allee effects, to understand how these effects may affect their populations and how their management (exploitation or conservation) must take these effects into account if it is to be successful. We will attempt to demonstrate this importance in six chapters.

This first chapter is aimed at introducing the concepts and giving you the desire to finish the book. If you read these lines, we will at least be optimistic that you will finish this chapter, which is already nice.

Chapter 2 considers a wide range of mechanisms which can potentially lead to Allee effects, from pollination to mate finding to cooperative hunting. We are sure you noticed that this first chapter lacks many concrete examples. This will be amended in Chapter 2. For each potential mechanism we will ask three questions: (i) *How does it work?* (ii) *Is there evidence that it creates a component Allee effects in real populations?* and (iii) *Is there evidence that the component Allee effect translates into a demographic Allee effect?*

Much of what we know about the Allee effects comes from mathematical models and Chapter 3 aims to present this field to the reader, concisely and from several complementary perspectives. In particular, we show what types of model are available, how they can be used, what are their advantages, disadvantages and underlying assumptions and, most importantly, how they have contributed to our understanding of the consequences of Allee effects for population and community dynamics. We present first single-species models, then models involving two or more interacting species. Because, although a really powerful tool in ecology, mathematical modelling unfortunately remains a strong deterrent to many ecologists, we have regrouped all stray equations into the third chapter, and they will not be found elsewhere in the book.

Chapter 4 concerns Allee effects in relation to genetics and evolution. As we saw above, we consider that genetic factors can be Allee effect mechanisms *per se*. We also feel that a glimpse into the evolutionary processes is a good avenue for a better understanding of Allee effects. Populations that are naturally rare, constantly or regularly, could have evolved means of mitigating or even circumventing component Allee effects. Do these adaptations represent the 'ghost of Allee effects past'? On the contrary, would populations that have always been numerous and stable be in theory more susceptible to Allee effects if they were to be significantly reduced? How do Allee effects shape the evolution of life history traits? These are some of the issues addressed in this fourth chapter.

Chapter 5 will present the implications of the Allee effect for the applied activities related to ecology. We will see how Allee effects can play a role in the decline or extinction of affected populations, how exploitation is doubly concerned, as it can both trigger a latent Allee effect or even entirely create one, and how population management, either aimed at protecting populations (e.g. reintroductions or captive breeding) or at controlling them (e.g. biological invasions and biological control) will be constrained by Allee effects. We will also present managers with some concrete advice on how to detect Allee effects and for different scenarios where they can be confronted by Allee effects.

Chapter 6 will present the key conclusions of the book and propose future directions of thoughts and research. In this final chapter, we will also discuss some other exciting and novel points which relate to Allee effects but which are rarely discussed in this respect.

The whole is designed so that each chapter can be read independently if needed (with only occasional cross-references to other chapters), and one chapter or the other might be skipped without affecting too much the others. Obviously, if only one chapter was to be left unread in the entire book, it would be Chapter 1. Well, perhaps we should have mentioned that earlier.

2. *Mechanisms for Allee effects*

Density dependence does not arise in populations like a law of physics. Just as negative density dependence arises from competition, positive density dependence—an Allee effect—must also be driven by some mechanism. A component Allee effect can arise via any mechanism which creates positive density dependence in some component of fitness. This may or may not then have consequences at the population level (a demographic Allee effect). In this chapter we look at potential mechanisms which create component Allee effects, and ask whether there is any evidence that these component Allee effects lead to demographic Allee effects.

2.1. Introduction

As we saw in Chapter 1, an Allee effect occurs when there is a positive relationship between a component of fitness and population size or density. 'Fitness' is of course a complicated concept, broadly defined as the genetic contribution of an individual to the next generation (see any ecology textbook for a discussion). Very broadly, fitness has two main components; survival and reproduction—in order to contribute to the next generation an organism has to first survive and then reproduce (and continue surviving to reproduce further). Survival and reproduction then divide into numerous subcomponents; for example, plant reproduction can be divided into components such as number of flowers, pollination rate, seed set etc. Reproduction and survival are also intertwined and non-independent—an offspring's survival is a component of parent's reproductive success, after all, and a characteristic such as bright plumage may have implications for both reproduction (higher) and survival (lower). This is particularly true of social and cooperative species, which accrue multiple fitness benefits from large population size (more efficient acquisition of resources, higher reproductive output, higher adult and juvenile survival and sometimes inclusive fitness from relatedness to other group members).

Nonetheless, survival and reproduction make a convenient, if imperfect, means of dividing up potential mechanisms for Allee effects according to the component of fitness in which they act. We present Allee effects mechanisms using this reproduction/survival partition, and discuss the case of social and cooperative species separately. Positive density dependence can occasionally arise in growth too, and we consider a few examples.

Mechanisms can also be divided into 'specific' mechanisms which pertain only to a particular type of life history (e.g. sessile organisms) and 'general' mechanisms that arise from broader ecological processes (e.g. predation). Allee effects with 'general' mechanisms are potentially applicable to a wider range of taxa. They are also harder to predict, because the mechanism depends on ecological circumstances (e.g. presence of particular predators), and so may be present in some areas or time periods and absent in others. There are also mechanisms which act on population *size* and mechanisms which act on population *density*, or both. This means that Allee effects can arise in dense populations (if small), and in large populations (if sparse).

Numerous Allee effect mechanisms exist, and there are many exciting studies which demonstrate these mechanisms and their impacts on natural populations. However, before starting in on our review, we need to point out a few 'health warnings' which should be borne in mind when reviewing any study which purports to demonstrate an Allee effect. Firstly, while it is easy to dream up possible mechanisms for Allee effects, we cannot assume that because a mechanism exists to create an Allee effect in theory, an Allee effect will in practice occur. And even if a component Allee effect does occur, this may not translate into a demographic Allee effect. In this chapter we have tried to avoid an uncritical listing of all the possible mechanisms for Allee effects, since our aim is not to be an 'advocate for the Allee effect' and convince readers that it lurks under every rock. Rather, we have tried to review critically the evidence that each mechanism may occur in natural populations. For some mechanisms, the evidence is limited; this may mean the mechanism is unlikely in practice or may mean that the relevant work has not yet been done.

Secondly, our definition of an Allee effect relies on the notion of 'fitness'. Fitness is a factor which has to be evaluated over the entire lifespan of the individual organism in question, or at least (more realistically) over the course of an entire seasonal cycle. Data collected on a single component of fitness at a single point in time can be misleading; for example an increase in reproduction in one year may be balanced by a decrease in reproduction or survival later due to resource limitation.

Thirdly, comparisons between small and large populations may be (often are) affected by confounding factors such as habitat quality. Large or dense populations may occur in good quality habitat and small or sparse populations in poor quality habitat. Individuals in large populations may then have higher fitness, but this is not likely to be due to an Allee effect. The detection of component and demographic Allee effects, and the problems of confounding variables are discussed further in Chapter 5 and 6, but as with every branch of science, ecological studies that purport to demonstrate Allee effects should be read with a critical eye.

The general mechanisms discussed in this chapter are summarized in Table 2.1. Examples of demographic Allee effects are set out in Table 2.2. A few potential mechanisms are not discussed in this chapter: inbreeding and other genetic Allee effects are discussed in Chapter 4, the consequences of human exploitation are considered in Chapter 5 and demographic stochasticity as an Allee effect mechanism (or not) is discussed in Chapter 6. There may well be other mechanisms that we have not thought of. Other reviews of Allee effect mechanisms are provided by Dennis (1989), Fowler and Baker (1991), Saether *et al.* (1996), Wells *et al.* (1998), Stephens and Sutherland (1999, 2000), Courchamp *et al.* (1999a, 2000a), Liermann and Hilborn (2001), Peterson and Levitan (2001), Gascoigne and Lipcius (2004a), Levitan and McGovern (2005), Berec *et al.* (2007) and references therein.

2.2. **Reproductive mechanisms**

Reproductive Allee effects are the most studied and best understood. Well-known mechanisms for reproductive Allee effects include fertilization efficiency in sessile organisms, mate finding in mobile organisms and cooperative breeding.

2.2.1. **Fertilization efficiency in sessile organisms**

How it works
Sessile organisms live permanently attached to land or seabed—plants are an obvious example but many animals are also sessile during reproductive maturity: notably marine invertebrates such as sponges, corals, anemones, oysters etc. Others are not obligately sessile but nonetheless move around very little; this includes most bivalves and echinoderms and many polychaete worms, as well as others; we might term these 'semi-sessile'. This mode of life requires individuals to reproduce without in most cases being able to come into direct contact with conspecifics. They therefore have to rely on the transfer of gametes through the surrounding medium (water or air). In animals this type of reproduction is termed 'broadcast spawning'; in plants, 'pollination'. In essence the Allee effect arises out of the physics of diffusion, which dictates that the further the gametes 'cloud' travels the more dilute it becomes, so individuals in sparse populations are likely to receive fewer gametes, making fertilization less efficient and successful sexual reproduction a potential problem at low density. Most work on this type of Allee effect has been done on plants, but there are also studies of Allee effects in broadcast spawning invertebrates, mainly semi-sessile echinoderms.

Table 2.1. Summary of mechanisms for component Allee effects with some examples. Adapted from Berec *et al*. (2007) and Stephens and Sutherland (2000). References given in text.

Mechanism	How it works	Examples
Component Allee effects in reproduction		
Broadcast spawning	Lower probability of sperm and egg meeting in water column at low population density	Several echinoderm species (Fig. 2.1), corals; most broadcast spawners?
Pollen limitation	Decreased pollinator visitation frequencies and lower probability of compatible pollen on pollinator at low plant density; lower probability of pollen grain finding stigma in wind-pollinated plants	Several self-incompatible insect-pollinated species (*Gentianella campestris*, *Clarkia* spp., Fig. 2.2), mast flowering trees, *Spartina alterniflora*
Mate finding	Harder to find a (compatible and receptive) mate at low population size or density	Cod, gypsy moth, Glanville fritillary butterfly, alpine marmot
Reproductive facilitation	Individuals less likely to reproduce if not perceiving others to reproduce, the situation more likely in small populations	Whiptail lizard, snail *Biomphalaria glabrata*; milk and queen conch, abalone, lemurs, many colonial seabirds
Sperm limitation	Females may not get enough sperm to fertilize eggs if males scarce during mating window	Blue crab, rock lobsters
Cooperative breeding	Breeding groups less successful in producing and/or rearing young when small	African wild dog
Component Allee effects in survival		
Environmental conditioning	Amelioration of environmental stress through large numbers via many specific mechanisms, such as thermoregulation (marmots), better microbiological environment (fruitflies) or overcoming host immune defences (parasites)	Alpine marmot, mussels, bark beetle, fruit fly, alpine plants, bobwhite quail, some parasites
Predator dilution (predator satiation or swamping)	As prey groups get smaller or prey populations sparser, individual prey vulnerability increases	Colonial seabirds, synchronously emerging insects, mast seeders, queen conch, island fox, caribou, American toad
Cooperative anti-predator behaviour	Prey groups less vigilant and/or less efficient in cooperative defence when small; fleeing small groups confuse predators less easily	Meerkat, desert bighorn sheep and other ungulate herds, African wild dog, schooling fish, lapwings
Human exploitation	See Section 5.2	See Section 5.2
Component Allee effects in reproduction and/or survival		
Foraging efficiency	Ability to locate food, kill prey or overrule kleptoparasites decline in small foraging groups and can, in turn, reduce individual survival and/or fecundity	African wild dog, black-browed albatross, passenger pigeon
Cultivation effect	Fewer adult fish imply higher juvenile mortality; fewer adult urchins worsen settlement success and feeding conditions of their young, and lessen protection from predation	Cod, many freshwater fish species, red sea urchin
Genetic Allee effects		
Genetic drift and inbreeding	See Section 4.1	See Section 4.1

Table 2.2. Examples of probable demographic Allee effect, with associated mechanism.

Species	Mechanism	Evidence	Human impacts?	Refs
Banksia goodii	Pollination failure	Reproductive failure of smallest populations	Yes—habitat loss and fragmentation	Lamont et al. (1993)
Clarkia concinna	Pollination failure	Reproductive failure of smallest populations if isolated	No	Groom (1998)
Spartina alterniflora	Pollination failure	Reduced (but still positive) per capita population growth rate at low density in front of invasion	Yes—invasive species	Davis et al. (2004), Taylor et al. (2004)
Gentianella campestris	Pollination failure	Threshold in extinction probability in non-selfing populations	Yes—extinction and reintroduction	Lennartsson (2002)
Glanville fritillary *Melitaea cinxia*	Mate finding	Hump-shaped relationship between per capita population growth rate and population size	Yes—habitat loss and fragmentation	Kuussaari et al. (1998)
Suricate *Suricata suricatta*	Predation	Higher survival in larger groups for adults; higher survival in larger groups for juveniles in presence of high predator densities	No (reduction of predators in ranchland mitigates Allee effect)	Clutton-Brock et al. (1999)
Atlantic cod *Gadus morhua*	Mate finding and/or cultivation effect	Per capita population growth rate ~zero in population reduced by overfishing	Yes—overfishing	Rowe et al. (2004), Swain and Sinclair (2000), Walters and Kitchell (2001)
Quokka *Setonix brachyurus*, black-footed rock wallaby *Petrogale lateralis*	Predation by foxes and cats	Threshold population size below which reintroduced populations go extinct	Yes—extinction by introduced predators; reintroduction	Sinclair et al. (1998)
Eastern barred bandicoot *Perameles gunnii*	Predation by foxes and cats	Increasingly negative per capita population growth rate as population shrinks	Yes—extinction by introduced predators; reintroduction	Sinclair et al. (1998)
California island fox *Urocyon littoralis*	Predation by golden eagles	Hump-shaped relationship between population growth rate and population size	Yes—predator facilitated by introduced prey	Angulo et al. (2007)

Species	Mechanism	Description	Allee effect	Reference
Caribou *Rangifer tarandus caribou*	Predation by wolves, bears and cougar	Higher per capita population growth rate in denser populations, threshold of ~0.3 animals per km^2	Yes—population fragmentation	Wittmer et al. (2005)
Red-backed voles *Clethrionomys gapperi*	Predation?	Threshold in habitat occupancy for lower quality habitat of ~13 voles ha^{-1}; positive density dependence in fitness according to ideal free distribution theory	No	Morris (2002)
Roesel's bush cricket *Metrioptera roeseli*	Unknown: not mate-finding	Persistence of introduced populations depends on propagule size: threshold ~16 animals	No—experimental study	Berggren (2001), Kindvall et al. (1998)
Hutton's and sooty shearwater	Predation by pigs and stoats	Threshold colony size below which colony goes extinct	Yes—introduced predators	Cuthbert (2002)
Thick-billed murre *Uria lomvia*	Predation by gulls	Large colonies stable, small colonies declining	No	Gilchrist (1999)
Lesser kestrel *Falco naumanni*	Predation by foxes and rats	Higher predation mortality in small colonies, emigration rates suggest higher overall fitness in large colonies	Yes—species threatened and in decline	Serrano et al. (2005)
Aphids *Aphis varians*	Predation by ladybirds	Positive relationship between colony size and per capita population growth rate	No	Turchin and Kareiva (1989)
Gypsy moth *Lymantria dispar*	Mate finding	Threshold colony size below which colony goes extinct	Yes—introduced species	Tcheslavskaia et al. (2002), Liebhold and Bascompte (2003)
Crown-of-thorns starfish *Acanthaster planci*	Release from predation causes outbreaks	Strong relationship between frequency and intensity of outbreaks and number of large fish predators	Yes—fishing pressure on predators	Dulvy et al. (2004)
Queen conch *Strombus gigas*	Predator dilution and reproductive failure	~100 % mortality of juveniles away from aggregation; reduced reproduction at low density; collapse and failure to recover in many areas	Yes—heavy fishing pressure	Stoner and Ray (1993), Ray and Stoner (1994), Marshall 1992, Berg and Olsen (1989)

Figure 2.1.   Sea urchins in a cloud of sperm. As the density of adults gets lower, species that reproduce by releasing their gametes into the environment can have lower reproductive success. Photo: Stuart Westmorland.

Examples

In both plants and invertebrates, fertilization efficiency is often positively related to population density, and a series of studies demonstrating this relationship are summarized in Figures 2.2 and 2.3. Note that in Fig. 2.2 (invertebrates) the x-axis is an (inverse) function of population *density* (nearest neighbour distance), while in Fig. 2.3 (plants) it is a function of population *size*. This reflects the fact that most work on pollen limitation in plants has been done on plants pollinated by some animal vector (usually insects). These actively seek out flowers, complicating the issue of dilution by diffusion. Instead, pollen limitation in these plants is driven by the fact that most pollinators move relatively short distances and are less likely to find, visit and spend time in a small patch relative to a large patch (Sih and Baltus 1987, Wilcock and Neiland 2002, Ashman *et al.* 2004). Also, a generalist pollinator (the majority) will visit plants of other species—and a higher proportion of these when the species of interest is at low density. Thus individuals in low density populations receive less pollen and a lower proportion of conspecific pollen than those in dense populations (Ashman *et al.* 2004, Wagenius 2006). Wind pollination is more directly analogous to broadcast spawning, and pollen limitation has been recorded in a wind-pollinated plant (cordgrass), where small, isolated patches of clones set little seed (Davis *et al.* 2004). In European beech (*Fagus sylvatica*), trees in smaller habitat patches set less seed (Nilsson

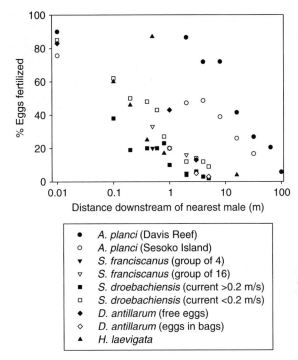

Figure 2.2. Fertilization efficiency in broadcast spawning invertebrates (four echinoderms, one gastropod), showing a general trend of exponential decline in the proportion of eggs fertilised with increasing nearest neighbour distance. Data from Babcock *et al.* (1994) (crown of thorns starfish, *Acanthaster planci*); Levitan *et al.* (1992) (red sea urchin, *Strongylocentrotus franciscanus*); Pennington (1985) (green sea urchin, *S. droebachiensis*); Levitan (1991) (Caribbean long-spined sea urchin, *Diadema antillarum*); Babcock and Keesing (1999) (green-lip abalone, *Haliotis laevigata*).

and Wästljung 1987), and seed set in the blue oak (*Quercus douglasii*) may also be limited by the number of neighbours in a ~60m radius (Knapp *et al.* 2001). High seed set in larger populations may not, however, always be a consequence of pollination. In the pale swallow-wort, *Vincetoxicum rossicum* (an invasive vine), larger populations had higher seed set as a consequence of the suppression of other plant species, and hence reduced interspecific competition in large patches (Cappuccino 2004). This is of course still an Allee effect, but with a different mechanism.

Despite a strong and intuitive mechanism for component Allee effects in these species, making the connection with demographic Allee effects is difficult, as illustrated in Box 2.1. There are some examples of demographic Allee effects via fertilization efficiency, although the evidence is mainly circumstantial. In the rare Australian endemic protea, *Banksia goodii*, reproduction in small isolated

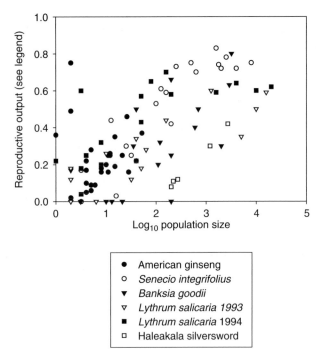

Figure 2.3. Reproductive output of flowering plants as a function of population size. American ginseng (*Panax quinquefolius*): green fruits per flower vs. number of individuals flowering (Hackney and McGraw 2001); *Senecio integrifolius*: proportion of seeds set vs. number of individuals flowering (Widén 1993); *Banksia goodii*: proportion of plants fertile vs. population size in m² (Lamont *et al.* 1993); *Lythrum salicaria*: seeds per flower/100 vs. number of individuals flowering (Agren 1996); Haleakala silversword (*Argyroxiphium sandwicense macrocephalum*): proportion of seeds set vs. number of individuals flowering synchronously (Forsyth 2003).

populations was found to be rare, and five of the nine smallest populations produced no fertile cones in ten years. The authors measured the most likely confounding variables (soil properties and under- and overstory vegetation cover) and found no relationship with *Banksia* population size or cone production (Lamont *et al.* 1993). Inbreeding might, however, provide an alternative mechanism. By contrast, in a study of the annual herb *Clarkia concinna*, the size and isolation of populations were manipulated experimentally. Again, reproduction in small, isolated populations was rare, and such patches were more likely to go extinct (Groom 1998). Neither of these studies could make a direct causal connection between population size, low seed production and extinction, however.

An alternative approach is to combine field data with population modelling; again this will not give a definitive causal connection between population size and population growth rate, but it can be nonetheless convincing. In *Spartina*

Box 2.1. **From component to demographic Allee effects? Fitness trade-offs**

For a demographic Allee effect to occur, overall fitness must increase with population size or density. In reality, even where there is an Allee effect mechanism in some fitness component, other fitness components are likely to be negatively density dependent, so the overall relationship between fitness and size or density depends on a series of trade-offs. These are particularly well illustrated with some examples from plant pollination and broadcast spawners.

Plants which have high rates of seed set (i.e. close to 100% fertilization efficiency) can suffer reduced growth and reproduction the following year. In one study, this cost of high seed set meant that populations with low fertilization efficiency had the same population growth rate overall as populations with high fertilization efficiency (Ehrlen 1992, Ehrlen and Eriksson 1995). Other trade-offs may be introduced by flower morphology. Plants can have multiple flowers open simultaneously ('display size'). An empirical test in the foxglove, *Digitalis purpurea,* showed a suite of complex interactions between density and display size, with the proportion of flowers visited increasing with increasing density for small display size but decreasing with increasing density for large display size; i.e. a component Allee effect was only present when display size was small (Grindeland *et al.* 2005).

Pollen limitation could also in theory trade off with reduced intraspecific competition at low density, but we have not come across a study which has addressed this hypothesis in plants. In sea urchins, however, this hypothesis has been tested. Echinoderms can pull off a variety of neat physiological tricks, including the ability to shrink their body mass when times get hard (Levitan 1991). Individuals in sparse populations are thus generally larger, because competition for resources is lower. In the Caribbean long-spined sea urchin, *Diadema antillarum,* increased gamete production by larger individuals at low density seems to offset the reduction in fertilization efficiency, so that individual reproductive output remains broadly similar across a wide range of densities (Levitan 1991). This study demonstrates the flaw in our slightly simplistic definition of reproduction as 'a component of fitness'. In fact, reproduction is itself made up of several components of fitness, of which fertilization efficiency is one and gamete production is another. Thus in *Diadema* there is a component Allee effect in fertilization efficiency which is offset by negative density dependence in total gamete production.

alterniflora, pollen limitation in sparsely populated areas was likely according to models to reduce the per capita population growth rate although it still remained positive at all sites (a weak demographic Allee effect) (Davis *et al.* 2004, Taylor *et al.* 2004). The same approach was used on reintroduced populations of the field gentian (*Gentianella campestris*), a herb of rare low-nutrient grasslands (Lennartsson 2002). Some individuals of *G. campestris* are self-fertile while others are not, and the trait is heritable. Non-selfing populations showed a threshold patch size below which seed set fell dramatically, and demographic models suggested that these populations had high extinction probabilities. In self-compatible populations, no such critical thresholds were predicted. A pollen-limitation Allee effect may thus be a driver for the evolution of self-fertile strains within some species such as *G. campestris* and the Californian annual *Clarkia xantiana* where self-fertile populations occur outside the range of three congeners which share pollinators, enhancing the size of the patch from the pollinator's point of view, In this example, 'population size' should actually be interpreted as the sum of several sympatric populations of different species (see also Section 4.1.5). Note, however, that these studies (or at least, our interpretation of them) conflates habitat patch size and plant population size; see Box 2.2.

In broadcast spawners the evidence for demographic Allee effects is likewise limited. Aronson and Precht (2001) suggest that they might be implicated in the failure of the corals *Acropora palmata* and *A. cervicornis* to recover from disease outbreaks and bleaching. Acroporids broadcast both eggs and sperm, and have largely been replaced on Caribbean reefs by *Agaricia* and *Porites* species, in which eggs are retained or 'brooded' by the females during and after fertilization, reducing sperm limitation. There is broader evidence for this idea in the fossil record, where comparisons of related taxa show that those with brooding larvae have suffered lower extinction rates in times of stress, implying that they can recover more easily from episodes of reduced population density or size (Wray 1995). However, this is far from being strong evidence of demographic Allee effects in these species; brooding species may just have a higher intrinsic per capita population growth rate or be more resilient to environmental change for some other reason.

Finally, plants and broadcast spawning invertebrates are clearly not completely comparable in their reproductive mode, and it is instructive to consider differences as well as similarities (see Box 2.3).

2.2.2. Mate finding

How it works

This mechanism is probably the most well known of all types of Allee effect, sometimes even being incorporated into the definition of the term. The basic

---

Box 2.2. **Habitat patch size vs. population size in plant populations**

In several examples, we use habitat patch size as a proxy for plant population size (note that this is our interpretation, not that of the authors concerned). In reality, the two are not the same thing, and we must be careful when using habitat patch size as a proxy measure of population size.

Plant population size and habitat patch size are likely to be correlated in a fragmented landscape, but there is no absolute guarantee (e.g. see Kolb and Lindhorst 2006). Pollinators are often generalists, and in this case will respond to the availability of a range of plant species, rather than just the species of interest to us. A small or sparse population may thus not suffer a reduction in pollinator visits if it is in a large habitat patch surrounded by other species attractive to pollinators. It may, however, still suffer a reduction in pollination rate since the majority of pollen will not come from conspecifics. Likewise, a dense population in a small habitat patch may be less attractive to pollinators because of a scarcity of other attractive species. Overall, pollination rates will probably be greatest for a large population in a large fragment, lowest in a small population in a small fragment and intermediate for large populations in small fragments or vice versa. As ever the reader is warned to look carefully at the specific variables used in each study.

---

theory is simple: at low density, individuals will not always be able to find a suitable, receptive mate, and their reproductive output will decrease accordingly.

There are a number of models which can be used to quantify the increase in mate searching time or the decrease in female mating rate with decreasing density (Table 3.2, Dennis 1989, Liermann and Hilborn 2001). They predict (unsurprisingly) that a mate-finding Allee effect is most likely in species which are mobile but have limited dispersal ability, in populations which are dispersed in space rather than aggregated (Dobson and Lyles 1989) and where individuals are only reproductive intermittently and asynchronously (Calabrese and Fagan 2004). Other than these 'pointers', and unlike with pollination or broadcast spawning, it is difficult to predict *a priori* whether this mechanism will exist in a given population.

Examples
Mate-finding Allee effects have been proposed in species on land from sheep ticks to condors and in the sea from zooplankton to whales (e.g. see reviews in

---

Box 2.3.   **Differences between fertilization in plants and aquatic invertebrates**

We have seen that there are many similarities in reproductive dynamics between plants and broadcast spawning invertebrates, including the possibility of Allee effects. However, there are also significant differences, and it is interesting to consider why this might be. For example, plants do not disperse both male and female gametes, as is common (although by no means universal) in broadcast spawners. More importantly, pollination of plants via an animal vector provides a highly evolved and sophisticated mechanism to facilitate reproduction and outcrossing even at low density which has never evolved in invertebrates, despite the fact that these invertebrate taxa have an evolutionary history which is many times longer than that of flowering plants.

The answer to this apparent paradox may lie in the physics of air vs. water. Wind speeds are higher than currents (1 ms^{-1} is a light breeze but a very strong current); this faster flow in air dilutes gametes faster and makes gamete capture more difficult. The higher density of water means that sinking in water happens much more slowly than falling in air. Thus this mode of reproduction is intrinsically more problematic on land than in water, and natural selection in plants to overcome problems associated with fertilization must therefore have been stronger than in aquatic invertebrates. This might also explain why a sessile life history is uncommon for land animals, and why the dispersal stage of terrestrial animals (insects, for example) is more likely to be the adult (reproductive) stage, while in marine animals it is more likely to be a larval stage. The evolutionary consequences of Allee effects are discussed in Chapter 4.

Dennis 1989, Fowler and Baker 1991, Wells *et al.* 1998 and Liermann and Hilborn 2001). For whales, they are a particularly alluring idea since it is not intuitive to us how whales can find each other across oceans. In blue whales, mating shortage was first proposed quite soon after WC Allee put forward his original ideas (Hamilton 1948), although there is still no evidence for or against the hypothesis in practice (Fowler and Baker 1991, Butterworth *et al.* 2002).

Moving down the size spectrum, we find stronger evidence for mate-finding Allee effects in invertebrates. Populations of the invasive gypsy moth have a positive relationship between the probability of a female being mated and the population density (Tcheslavskaia *et al.* 2002), leading to a demographic Allee effect (Liebhold and Bascompte 2003, Johnson *et al.* 2006), and holding out some hope of halting the advance of this damaging species across North America (see Section 5.2). Mate-finding Allee effects are also proposed in some species of copepods (Kiørboe 2006), and as the mechanism through which some pastures stay mysteriously free of sheep ticks (Milne 1950, cited in Liermann and Hilborn 2001). Mate-finding Allee effects do not necessarily require two separate sexes in the usual 'human' sense; the wheat pathogen, *Tilletia indica*, reproduces via encounters between sporidia of different mating strains, which are less frequent at low density (Garret and Bowden 2002). They might even occur in the malaria parasite within the body of humans and mosquitos (Pichon *et al.* 2000). Allee-type effects in parasites are discussed in Section 3.6.3.

Some of the most striking elements of animal behaviour are adaptations for finding mates, including calls and song, displays, odour and pheromone marking, reproductive aggregations and so on. A mate-finding Allee effect may arise if these mate-finding behaviours are themselves disrupted at low density. For example, heavy fishing on spawning aggregations of reef fish results in knowledge of spawning sites and migration routes being lost from the population, as well as disruption of spawning behaviour via the removal of dominant males (Sadovy 2001), although Allee effects have not been specifically demonstrated in these populations.

The role of dispersal in mate-finding Allee effects

Dispersal or movement rates are key to creating (or avoiding) mate-finding Allee effects. An increase in movement rates within low density populations

Figure 2.4. The Glanville fritillary butterfly, *Melitaea cinxia*, one of the first species in which a mate-finding Allee effect has been demonstrated. Photo: Marjo Saastamoinen.

can effectively counteract mate-finding Allee effects, as in the bush cricket, *Metrioptera roeseli* (Kindvall *et al.* 1998). If, however, individuals are more likely to disperse away from small or low density populations in search of mates (or more generally in search of a higher fitness habitat), this can exacerbate the Allee effect by reducing the per capita population growth rate of these populations still further. In the Glanville fritillary butterfly (*Melitaea cinxia*; an endangered species of northern European dry meadows), a higher proportion of males emigrate out of small populations in search of mates, and a lower proportion of females are thus mated. When populations were isolated their per capita growth rate had a domed relationship with population size, with a maximum growth rate in intermediate size populations (Kuussaari *et al.* 1998). As for plants (Groom 1998) the interaction of population size and isolation is important.

2.2.3. **Sperm limitation**

How it works

A related issue is sperm limitation, which in mobile organisms is a question of mate finding more than fertilization efficiency. In order to maximize her reproductive output, a female must find enough males, or a large enough male, to provide enough sperm to fertilize all her eggs. In sparse populations, she may not encounter any males, or enough males, or may not be able to choose the optimum size of male.

Examples

Evidence for sperm limitation in natural populations of mobile species is somewhat limited. Studies have mainly focused on exploited populations, and must be treated with caution as examples of Allee effects, because sperm limitation may arise due to the exploitation itself, as well as due to small population size. This occurs when the fishery targets large males in particular. Females in heavily exploited populations of blue crab (*Callinectes sapidus*), Caribbean spiny lobster (*Panulirus argus*) and New Zealand rock lobster (*Jasus edwardsii*), for example, are frequently sperm-limited because the fishery has reduced the abundance of large males, reducing mate choice and reproductive output of individual females (Macdiarmid and Butler 1999, Hines *et al.* 2003, Carver *et al.* 2005). The distinction between sperm limitation due to low density and sperm limitation due to lack of mate choice (lack of large males) is important, because in the former case, the rate of recovery of the population if the fishery were stopped would be low (an Allee effect), while in the latter case, the rate of recovery would be rapid once enough of the small males had grown to become large males. In practice, however, it is probably impossible to distinguish the two effects. In fact, other life history components in these species make demographic Allee effects unlikely or difficult to demonstrate; blue crabs are highly cannibalistic (Zmora *et al.* 2005) so that impacts of low density (sperm limitation) trade off with increased survival, while rock and spiny lobster larvae spend several months in the plankton, effectively decoupling local reproduction and recruitment (Phillips 2006).

2.2.4. **Reproductive facilitation**

How it works

In some species, individuals who do not encounter enough conspecifics do not become reproductive, not directly because they cannot find a mate, but rather because they need the presence of conspecifics to come into physiological condition to reproduce. The specific mechanism may vary; individuals may require stimuli through exposure to conspecifics, potential mates, courtship or mating behaviour or perhaps some other related factor—in most cases the precise mechanism is unknown.

Examples

One of W.C. Allee's best-known experiments, in laboratory populations of the flour beetle, showed that a hump-shaped relationship between per capita reproductive rate and density arise because the beetles need to encounter a certain density of conspecifics or mates to come into reproductive condition (Allee 1941, 1949). Reproductive facilitation may be important in queen and milk conch (*Strombus gigas* and *S. costatus*), large gastropods native to the sub-tropical western Atlantic and Caribbean. Queen conch have been heavily exploited throughout their range and in Bermuda and Florida populations have crashed and are showing limited signs of recovery (Berg and Olsen 1989). A survey of extensive deep-water populations of queen conch during the reproductive season showed that below a critical density of ~50 animals per hectare, there was no reproductive activity (Stoner and Ray-Culp 2000). A translocation experiment in shallow-water populations also suggested an Allee effect (Gascoigne and Lipcius 2004c). This may be a straightforward problem of mate finding, but reproductive facilitation may occur, because conch females engaged in egg-laying are ~8 times more likely to copulate than those not laying eggs (Appeldoorn 1988). Abalone (*Haliotis* spp.) is another large gastropod which may require reproductive stimulation by conspecifics. Abalones are broadcast spawners, but increase fertilization rates by aggregating to spawn. As density declines, a decreasing proportion of reproductive adults participate in reproductive aggregations, reducing per capita reproductive output (Shepherd and Brown 1993).

Animals which can reproduce asexually may nevertheless have higher reproductive output in the presence of other individuals (Thomas and Benjamin 1973). In self-fertile snails (*Biomphalaria glabrata*) and parthenogenetic female lizards (*Cnemidophorus uniparens*), individuals housed in isolation produce fewer offspring than individuals housed in groups, apparently because of exposure to courtship behaviour, although they do not actually mate (Crews *et al*. 1986, Vernon 1995). It's not clear what effect this might have in nature (if any) but it is interesting and counterintuitive that even species which are self-fertile have the potential to suffer from reproductive Allee effects—compare with the results for plants discussed above. This kind of facilitation also seems to occur in lemurs and flamingoes in captivity (Stevens and Pickett 1994, Hearn *et al*. 1996, Studer-Thiersch 2000; see Section 5.1.3) and may be important in some colonial seabirds.

2.2.5. **Female choice and reproductive investment**

How it works

Sexual selection—that is, selection pressure on various characteristics via mate choice—plays an important role in the ecology as well as the evolution of many species. Mate choice is usually exercised by females, and can take various forms, depending on the benefits which females accrue from mating with more attractive males—these benefits can be direct (male parental care) or indirect (more genetically fit or attractive male offspring) (Møller and Thornhill 1998). In small or low-density populations where females cannot choose between males, or where choice is limited to males which are not particularly attractive, females may choose not to mate, or may invest less in reproduction and offspring, leading to lower reproductive success (Møller and Legendre 2001).

Examples

In models, this scenario can lead to a demographic Allee effect (Møller and Legendre 2001), but its importance in nature is less clear. The idea is difficult to test in natural populations, although there is good evidence in birds and insects that females mated with less preferred males have lower reproductive output (reviewed in Møller and Legendre 2001). Empirical evidence for Allee effects through this mechanism mainly comes from populations held in captivity, where population size is reduced to one or a few pairs—a level likely to lead to extinction in the wild even without Allee effects. This mechanism clearly does create headaches for captive breeding programmes in, for example, big cats, giant panda, gorillas and so on (Møller and Legendre 2001).

2.3. **Mechanisms related to survival**

2.3.1. **Environmental conditioning**

How it works

One of W.C. Allee's early ideas (1941) was that species may condition their environment to make it more favourable. Of course, 'environment' is a tricky term, since it can mean numerous different things; physical (temperature, wind, turbulence etc.), chemical (e.g. oxygen, toxins, hormones) and biological (e.g. competitors and predators). W.C. Allee had the physical and particularly the chemical environment in mind, and carried out numerous experiments showing that larger groups of freshwater crustaceans and fish were better able to 'condition' their water by removing toxins and secreting beneficial chemicals (exactly what chemicals is never completely clear, and was probably more difficult to assess in those days). In this section, we also consider the physical and chemical environment, as well as part of the biotic environment, to include microorganisms. Interactions with competitors and predators are not included in our definition of 'environment': predators are a key source of Allee effects and are considered in detail in Section 2.3.2 below.

The general idea of this mechanism is that if individuals improve their environment in some way, such that others benefit from it, individuals in larger groups will be living in a better environment. This may mean that the temperature is maintained at some optimum, they are protected from weather or physical stress, or there are lower concentrations of toxic chemicals and harmful microorganisms and/or higher concentrations of beneficial ones. This Allee effect mechanism can have interesting effects at the landscape level, see Box 2.4.

Examples

An obvious component of the physical environment is temperature, and there are several examples of species which need to overwinter in groups for protection against cold, including Alpine marmots (Stephens *et al.* 2002a), bobwhite quail (Allee *et al.* 1949) and monarch butterflies (probably a secondary benefit of aggregation for predator dilution or mating enhancement in this case—an interesting example of multiple potential benefits from high density; Calvert *et al.* 1979, Wells *et al.* 1998). Soil-dwelling isopods (woodlice) forage independently but group together when sheltering to reduce water loss and oxygen consumption; individuals at low density have poor survival (Brockett and Hassall 2005).

WC Allee's ideas on 'environmental conditioning' have recently been revived in a new form as a possible explanation for swarming behaviour in zooplankton such as mysid shrimp and krill (although predation may also play a role, see below). In experiments, individual mysids in swarms of >50 individuals consumed significantly less energy than individuals in groups of <10. Individuals may use the hydrodynamic wake shed by the others, or the whole group, to optimize feeding, reduce sinking rates and reduce the energetic costs of diel vertical migration (Ritz 2000).

Environmental conditioning also occurs in species that have symbiotic relationships with microorganisms such as mycorrhiza (plants) and yeasts (fruitflies) (Stephens and Sutherland 2000, Wertheim *et al.* 2002, Rohlfs and Hoffmeister

2003) or compete with microorganisms (fruitfly larvae/fungi) (Rohlfs *et al.* 2005). Some insects which secrete chemicals to modify or pre-digest vegetable food (many sap-feeders, grain-borers and tuber-feeders) require high densities for efficient feeding (Liermann and Hilborn 2001, Sakuratani *et al.* 2001).

In some insects, females deliberately aggregate their eggs, sometimes with those of other females. This includes the wasp *Micropletis rufiventris*, a parasitoid of the moth *Spodoptera littoralis* (agricultural pests known as Egyptian army-worms or Mediterranean climbing cutworms). The proportion of parasitoid eggs which develop through the larval stage to the final instar increases with the number of parasitoid eggs inside the larvae (Fig. 2.5). The same phenomenon apparently occurs with other parasitoid wasps (e.g. Perez-Lachaud and Hardy 2001). The way that this mechanism functions is not known, but might relate to the endocrine environment of the host, and may only apply to parasitoid species which feed selectively on host tissues, rather than consume the entire host organism (Hegazi and Khafagi 2005).

This sort of component Allee effect mechanism in parasites and parasitoids might be quite common (Regoes *et al.* 2002), but other components of fitness are likely to be negatively density dependent, so the probability of a demographic Allee effect is hard to assess. In *M. rufiventris*, while multiple larvae develop most successfully inside the host, usually only one or at most a small number emerge successfully (Hegazi and Khafagi 2005). In *Metarhizium anisopliae*, a fungal parasite of leaf-cutter ants, there is a density threshold for infection of a host (a component

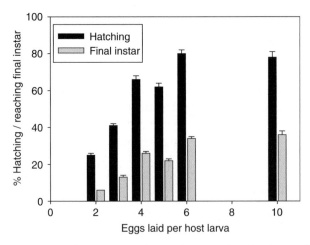

Figure 2.5. Development of *Micropletis rufiventris* larvae inside *Spodoptera littoralis* larvae. The parasitoids hatch and develop better in larger groups, perhaps because they are better able to overcome host immune defences (Hegazi and Khafagi 2005).

Box 2.4.　**Positive feedbacks, population boundaries and self-organized patterning**

There has been much recent work on interactions between neighbouring individuals along gradients of environmental stress. In some physically harsh environments (e.g. the intertidal, high-altitude and semi-arid zones, high energy marine environments) attached organisms such as mussels, barnacles, salt-marsh grasses and alpine plants can facilitate the growth and survival of their neighbours through various mechanisms, but under less harsh conditions they compete (see review in Gascoigne *et al.* 2005). This results in abrupt population boundaries in harsh environments because individuals cannot persist below a threshold density, as seen, for example, in alpine tree lines. Such transitions require some mechanism for positive feedback in density, which may vary from place to place, but could be, for example, that tree seedlings cannot become established except in the presence of mature trees which provide protection from wind, trap snow and change soil characteristics (Wilson and Agnew 1992, Alftine and Malanson 2004).

This type of Allee effect, interacting with negative density dependence, can explain some striking patterns seen in nature in environmentally harsh or

Figure 2.6.　A patterned bed of mussels (*Mytilus edulis*) of soft sediment. Photo: Joanna Gascoigne.

Box 2.4.   (*Continued*)

---

stressful habitats such as semi-deserts, peat bogs and sand flats. Despite the uniform nature of the underlying substratum, attached biota such as plants or invertebrates can be distributed in a highly patchy way. Furthermore, this patchiness can come in the form of patterns such as lines, 'tiger-stripes' or islands, with a characteristic wavelength. Empirical and theoretical work (e.g. Klausmeier 1999, Rietkerk *et al.* 2002, 2004, van de Koppel *et al.* 2001, 2005, Gascoigne *et al.* 2005 and references therein) has concluded that this 'self-organized' patterning can arise when both positive and negative relationships between fitness and density operate simultaneously, but at different spatial or temporal scales. For example, in semi-arid vegetation, plants facilitate the growth and survival of those nearby (a component Allee effect) by shading the soil (reducing evaporation) and by breaking up the soil (improving water infiltration). At the same time, however, they compete for water over a larger scale, meaning that plants on the edge of a large patch will not survive (Rietkerk *et al.* 2002). Likewise in mussel beds on soft sediment, a component Allee effect occurs because the byssal thread attachments of mussels on their neighbours protect mussels from being swept away during storms, but at the same time mussels in dense patches compete strongly for food (Gascoigne *et al.* 2005). These processes may result in a demographic Allee effect in the sense that if cover of biota were lost, it would be hard for a population to re-establish, even if a source of propagules were available—a possible cause of desertification, which can occur if plant biomass is removed by grazing (Schlesinger *et al.* 1990, Srivastava and Jeffries 1996, Rietkerk *et al.* 2004). Wilson and Agnew (1992) provide a comprehensive review of positive feedbacks in plant communities.

Allee effect), probably related to the host's immune response, but strong competition between parasites for resources within the host (Hughes *et al.* 2004).

2.3.2. **Predation**

How it works

Predation can generate an Allee effect in prey. This is clearly a very important and potentially wide-spread Allee effect mechanism, but unlike the other mechanisms set out in this chapter, it is not intuitive to understand, and thus merits a small amount of theoretical explanation. The process is explained in more detail in Section 3.2.1. Humans as predators are discussed in Section 5.2.

Predators respond behaviourally to the size and density of prey populations in various ways. Firstly, individual predators move around in search of denser areas or larger populations of prey ('aggregative response'). Predators can also eat more if more prey are available to eat ('functional response'), up to a limit. Predator populations too may respond to changes in prey population size or density, through changes in their own population size and growth rate ('numerical response'—note that this happens on a longer timescale than the functional and aggregative responses).

In simple terms, a decrease in prey density is likely to lead to a decrease both in predator numbers and predator consumption rate, but this decrease is not usually enough to balance the decrease in prey numbers (i.e. overall predator consumption of prey decline at a slower rate than prey numbers). This means that as prey density decreases, there are fewer prey individuals per predator attack, and thus each prey individual has a higher probability of being eaten. This process is sometimes called the 'dilution effect', because prey 'dilute' the probability of predation for each prey individual at high density. The dilution effect leads to a component Allee effect in predation mortality.

In natural systems there are, of course, a variety of mechanisms which stabilize predator–prey dynamics. The most obvious is that prey maintain high density or large population size via some behaviour such as aggregation or synchrony (see below). In other cases, predators actively avoid areas with low density prey populations, or switch to alternative prey species at low prey density (a 'type III functional response'), in which case low density can provide protection from predation (Gascoigne and Lipcius 2004a, see Section 3.2.1). This is common where prey use a predation-avoidance mechanism, such as crypsis, which is more effective at low density (Seitz *et al.* 2001). Spatial or temporal refuges from predation may also stabilize predator–prey dynamics (Gascoigne and Lipcius 2004a).

The protective effect of high density for prey via the dilution effect is distinct from any anti-predator benefits which arise from prey behaviour such as aggregation, temporal synchrony, group vigilance and group aggression towards predators. Below we discuss examples of each, organized by prey behaviour, from straightforward dilution, through prey aggregation in space or time, to more complex prey behaviours such as group vigilance and aggression.

Predation is the classic example of a 'general', extrinsic Allee effect mechanism, not limited to any particular life history, except to the extent that predation needs to be an important source of mortality. This makes the presence of a predation-driven Allee effect difficult to predict, and also means that it may occur in some small or sparse populations and not in others, or in a given population only at certain times.

Predator dilution without gregarious behaviour

An Allee effect via this mechanism is found in caribou (*Rangifer tarandus caribou*), where predation is the main source of mortality. Caribou are secondary prey of their predators (wolves, cougar and bears) with the main prey being other more abundant ungulate species including moose. In populations in British Columbia and Idaho, the per capita population growth rate of caribou decreased with decreasing population density, particularly at low density. Factors related to reproduction were ruled out, and the best hypothesis was that declines in sparse populations were related to a higher risk of predation mortality. The Allee threshold seems to be about 0.3 animals/km^2 (Fig. 2.7; Wittmer *et al.* 2005).

Another example is the island fox (*Urocyon littoralis*), a critically endangered carnivore endemic to the California Channel Islands already mentioned in Chapter 1. The fox provides a convenient system for looking at population size and predation, since different islands support fox populations of different sizes, with predators (golden eagles *Aquila chrysaetos*) only present on some islands. Golden eagles are the main predators in the system, and consume both foxes and feral pigs, with pigs as the main prey and foxes as secondary prey. The foxes themselves mainly eat small mammals, invertebrates, fruit etc. Before the introduction of pigs, eagles were only transient visitors to the Channel Islands, since fox populations by themselves did not provide enough prey to sustain a permanent population, but since the introduction of pigs eagles have established resident populations on the northern islands (Roemer *et al.* 2001). Fox populations declined precipitously in the 1990s in the islands where eagles are resident (Roemer *et al.* 2002). Fox survival and per capita

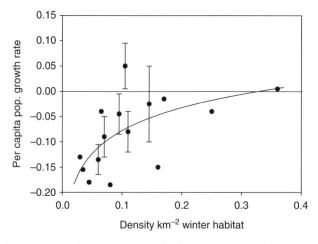

Figure 2.7.   Per capita population growth rate of 15 subpopulations of caribou as a function of density in their winter habitat. Predation was the main source of mortality in 11 subpopulations. From Wittmer *et al.* (2005).

population growth rate declines faster as fox density is reduced, revealing a demographic Allee effect in northern populations, which arises from increasing individual risk of eagle predation at low fox density (Figs 2.9 and 2.10; Angulo *et al.* 2007). Because eagle population dynamics do not depend on fox density, they can deplete the fox population without negative feedback, as long as there are sufficient pigs around to maintain a high density of eagles. As a result, the fewer foxes there are, the more each of them is likely (relatively) to be preyed upon. Ecologically, this is an interesting example because these Allee effects in the island fox are mediated by an indirect interaction with an introduced prey species (the pigs). It also posed a conservation conundrum because to save the island fox from extinction required removal of golden eagles, which are a protected species (Courchamp *et al.* 2003).

Prey aggregation in space: the 'selfish' herd

We are all familiar with large animal aggregations (herds of wildebeest, schools of fish, flocks of starlings, plagues of locusts etc.). Predator avoidance is one of the significant benefits of such aggregations (Hamilton 1971, Morton 1994, Reluga and Viscido 2005 and references therein), although they have other benefits, such as improved reproductive or foraging success (Krebs *et al.* 1972; see below for a discussion of social and cooperative species).

The idea of gregariousness as an adaptation to avoid predation was put forward many years ago, notably by WD Hamilton, who first suggested the concept of the 'selfish herd' (Hamilton 1971). The herd is 'selfish' because it forms through the

Figure 2.8. The island fox, *Urocyon littoralis,* is subject to a predator-driven Allee effect. Copyright: National Park Service US Gvt.

action of individuals trying to reduce their own risk of predation at the expense of other in the herd. Lacking knowledge about where the predator attack will come from, they can best minimize their own risk by staying close to other individuals, because this minimizes their 'domain of danger'—the area in space where they

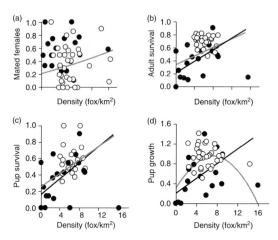

Figure 2.9. The relationship between various fitness components: (a) proportion of mated females, (b) adult survival, (c) pup survival, and the per capita population growth rate (d) with fox density in the northern islands (dark circles, with resident eagles) and the southern islands (open circles, without resident eagles). Statistically significant fits are shown for northern island populations only (black line) and for all populations (grey line). From Angulo *et al.* (2007).

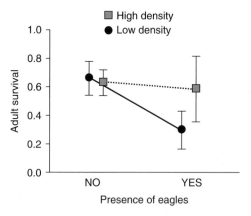

Figure 2.10. Fox populations were classified as high density (squares) and low density (circles). There is a component Allee effect in adult survival (significantly lower survival at low density) but only in the presence of the predator. From Angulo *et al.* (2007).

are likely to be the closest to a predator, if a predator attacks at a random point. If predator attack comes from outside the group, the advantage to herding is particularly great because only those on the outside have any domain of danger at all (Hamilton 1971, Morton *et al.* 1994, Reluga and Viscido 2005).

In theory this process should lead to a component Allee effect, because smaller groups have a higher ratio of edge to middle, and therefore a higher mean 'domain of danger' for their members. What evidence exists that this occurs in practice? An increase in survival in larger aggregations has been demonstrated in various arthropods (e.g. marine insects; Foster and Traherne 1981, aphids; Turchin and Kareiva 1989, bark beetles; Aukema and Raffa 2004, spiders; Avilés and Tufiño 1998 and monarch butterflies; Calvert *et al.* 1979). In juvenile queen conch, individuals outside the aggregations are highly unlikely to survive (Stoner and Ray 1993, Ray and Stoner 1994, Marshall 1992). The American pronghorn efficiently faces predators when in large groups, side by side, horned heads outside the circle. But in groups of less than 12–14 individuals, this protection mechanism becomes ineffective, the herd flees, and individuals are killed more easily (Allee *et al.* 1949). Seabird chicks of several species suffer less predation in larger colonies (reviewed in Liermann and Hilborn 2001). The mass 'arribadas' of olive and Kemp's ridley turtles on nesting beaches is also thought to be linked to predator dilution (Eckrich and Owens 1995). Evidence for demographic Allee effects in these examples is more limited, but in aphids, larger colonies have higher per capita population growth rates associated with lower per capita predation rates (Turchin and Kareiva 1989).

Some animals, most notably birds, aggregate together particularly for breeding. This colonial breeding is not the same as cooperative breeding (discussed below), where individuals other than the parents help to raise young; here we are simply talking about individual pairs raising chicks in close proximity to other pairs, without any inter-pair cooperation. The costs and benefits of coloniality have been studied most closely in seabird species, some 98% of which nest in colonies (Lack 1968, Rolland *et al.* 1998). Bird eggs and nestlings are very vulnerable to predation, particularly since adults have to leave the nest to find food, often for long periods of time. In colonies of Audouin's gull (*Larus audouinii*), female fecundity (chicks per female) increased with gull density, and particularly strongly with the ratio of predators to prey, suggesting a dilution effect (Oro *et al.* 2005). Note, however, that coloniality in seabirds may have evolved not primarily for predator protection but rather as a consequence of exploiting resources—schools of fish or invertebrates—which are (or were) abundant but are also very patchy and unpredictable in space and time (Danchin and Wagner 1997, Grünbaum and Veit 2003). This 'foraging' Allee effect is discussed further below. Gregariousness may also be a side effect of limited nesting sites such as cliffs and offshore islands, although birds which nest in more accessible sites (such as many gulls and terns) are still colonial, despite having cryptic eggs and young which on the face of it would be less visible away from high density, noisy and very obvious colonies (Lack 1968).

Whatever the main evolutionary driver for coloniality, individual bird survival often increases with colony size as a consequence of predator dilution. Cliff swallows (*Petrochelidon pyrrhonota*) in larger colonies had lower mortality from predation by great horned owls (Brown and Brown 2004). Colonies of Hutton's and sooty shearwaters (*Puffinus huttoni, P. griseus*) on the mainland of New Zealand suffer from predation by introduced pigs and stoats, with small colonies having higher rates of adult and chick mortality than larger colonies. In this case, predator dilution is likely to lead to a demographic Allee effect, since small colonies were declining (to extinction in some cases) due to heavy predation, while large colonies could persist (Cuthbert 2002). Likewise, in colonies of thick-billed murre (*Uria lomvia*, also called Brünnich's guillemot), predatory gulls could attack murres throughout small, sparse colonies, but are generally restricted to colony edges in large dense colonies (Gilchrist 1999; although in this case the small and large colonies were geographically far apart and may have differences other than size and density—the problem of confounding variables).

The lesser kestrel (*Falco naumanni*) is another bird species which often (although not always) breeds in colonies. Large colonies suffer significantly lower rates of predation on nestlings, and probably also on adults (Serrano *et al.* 2005; Fig. 2.11). Dispersal rates between colonies are higher from small to large colonies than from large to small colonies, suggesting that mean fitness is improved

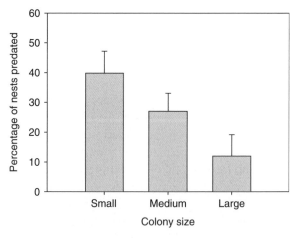

Figure 2.11.   Percentage of nests predated in small (1–3 pairs), medium (4–9 pairs) and large (>9 pairs) breeding colonies of the lesser kestrel *Falco naumanni*; means of six years of data from 1993–8. Adapted from Serrano *et al.* (2005).

in large colonies (Serrano *et al.* 2005). There is also a higher frequency of subordinate birds in small colonies (Serrano and Tella 2007), suggesting that they are a less favourable place to breed. There is thus potential for a demographic Allee effect in this species, but high rates of dispersal makes measurement of the population growth rate difficult, since colony size is determined as much by immigration and emigration as births and deaths.

Finally, it is interesting to note that flocking and herding species generally aggregate even in large populations and at high density (see the herd behaviour below). This suggests that the benefits of aggregation (in mitigating a component Allee effect) outweigh the costs (increased competition) across the whole range of population size and density, rather than just at small size or low density. Allee effects may thus affect animal behaviour and individual fitness at high as well as low density.

Herd 'behaviour'
Some species exhibit more complex herd 'behaviour' such as moving in apparent synchrony (schools of fish, flocks of starlings). It is hypothesized that this herd 'behaviour' enhances the benefits of aggregation by reducing the ability of predators to pick out individuals from the herd or school—the so-called 'confusion effect' (Tosh *et al.* 2006, Turesson and Broenmark 2004, Schradin 2000). The confusion effect has mainly been studied in piscivorous and planktivorous fish, and in some cases appears to operate (Botham and Krause 2005) while in other cases there is no evidence for it (Rickel and Genin 2006). In theory, the confusion effect should work better in larger groups, and thus enhance the straightforward 'dilution' benefit of group size. In practice it is difficult to separate the two.

Figure 2.12. Large swarm of red-billed queleas (*Quelea quelea*) returns to the communal roost at dusk. Copyright: Peter Johnson/CORBIS.

Prey aggregation in time

Prey may aggregate in time as well as space, on the basis that a synchronous emergence or migration of prey at high density will not allow predator populations time to respond and will 'swamp' their ability to feed, thus increasing mean individual survival from predation ('predator satiation')—a temporary dilution effect, in other words. The impact is reinforced if low prey density at other times keeps predator populations low, but it may also act together with aggregation in space. For example, colonially nesting seabirds synchronize egg laying (Lack 1968).

Aggregation may apply to seeds or propagules as well as directly to individuals. For instance, while trees themselves obviously cannot form temporary aggregations, they may practice 'mast seeding'. This is where all the individuals across a region (which can be continental in scale; Koenig and Knops 1998) flower and set seed simultaneously in certain years, while in other years almost none flower. Masting occurs in a variety of species, particularly oaks (mainly *Quercus* spp.) but also beeches (*Fagus* spp.), rowan (*Sorbus aucuparia*) and several species of tropical dipterocarps (Koenig and Ashley 2003, Koenig and Knops 2005) among others. Masting is sometimes irregular (e.g. dictated by weather conditions; Koenig and Knops 2005) and sometimes more regular (e.g. a two- or three-year cycle in rowan; Satake *et al.* 2004). In this latter case predator species can respond with their own adaptations to take advantage of a predicable, if highly variable, food source—via extended diapause in insect seed predators, for example (Maeto and Ozaki 2003). It seems likely that individuals of a masting species in low density or small populations (or individuals which flower out of synchrony) will suffer high rates of seed predation, although evidence is limited (Nilsson and Wästljung 1987, see below).

Masting has long been assumed to be an adaptation for predator satiation (e.g. Janzen 1971), but more recent work on the mechanisms behind masting suggests that it may be a consequence of pollen limitation (Nilsson and Wästljung 1987, Knapp *et al.* 2001, Satake and Iwasa 2002, Koenig and Ashley 2003, see also Forsyth 2003, Fig. 2.3). While the seed production of many masting tree species shows strong synchrony with weather patterns (e.g. spring temperature, Koenig and Knops 2005), modelling suggests that this by itself cannot generate such strong 'boom or bust' dynamics. However, if a positive feedback due to pollen limitation is included, such that flowering will only result in seed production when neighbouring trees are also flowering, strong spatial synchrony in seed production can be induced, either erratically or with a regular periodicity of two or more years (Satake and Iwasa 2002). This process is known as 'pollen coupling'. Pollen coupling can be regarded as a type of temporal component Allee effect, where individuals cannot reproduce unless the 'population' of other reproductive individuals is sufficiently dense. Generally, it is worth bearing in mind that the 'effective' population size from the point of view of reproduction may not always be as large as the absolute number of mature individuals; a point that is recognized, for example, by IUCN in its rules for assessing populations for their Red List status (see Section 5.1.4).

Predator satiation and pollen coupling probably operate together. In European beech, trees in smaller stands (patches) set less seed, and lose a higher proportion of it to invertebrate seed predators, although the relationship does not hold for vertebrate seed predators (Nilsson and Wästljung 1987). Three of the eight oak species in California (all masting) are suffering from a decline in regeneration rates throughout their range, which might arise through an Allee effect in conjunction with dramatic rates of habitat fragmentation and loss (estimated at about 100 km^2 per year in total). However, other explanations, such as reduced survival of seedlings due to overgrazing, changes in the fire regime and changes in other vegetation, may be more likely (Koenig and Ashley 2003, Walter Koenig, pers. comm.).

Other examples of predator satiation by synchrony in time include annual and periodical cicadas, which have long duration nymph stages underground, and then emerge synchronously as adults after several years (depending on the species). A predator satiation hypothesis seems the most likely explanation for this behaviour, since adult cicada individuals lack even basic anti-predator behaviour (Grant 2005). Emergence years often have a cycle with a period of a prime number (11, 13 or 17 years) and it is speculated that this has evolved to prevent predators from evolving a cycle with a shorter period which would coincide with the cicada emergence every time (as would occur if say, the predators had a cycle of period 2 or 4 years and the cicadas 8 or 16 years). Again, not much information is available on whether individuals in larger or denser swarms really do have reduced predator mortality. American toad (*Bufo americanus*) tadpoles also aggregate more strongly and metamorphose more synchronously in the presence of predators, at least in the lab (DeVito 2003).

The mass migrations of species such as wildebeest and salmon also act to satiate predators, alongside other ecological functions such as reproduction. In salmon, the risk of mortality from predation by birds and humans is lower where the migrating stock is larger (Wood 1987, Peterman 1980). However, so many factors are interacting to threaten wild salmon populations (dams, catchment change, overfishing, interbreeding with farm escapees...) that clarifying the role of this potential component Allee effect in the population dynamics is probably impossible.

Group vigilance and aggression
The protective effect of aggregation against predation can be enhanced by group anti-predator behaviour. Many species show greater individual predator vigilance behaviour when in small groups or at low density; e.g. desert bighorn sheep, pronghorn antelope, ibex, springbok and other ungulates (Mooring *et al.*

2004); marmots (Quenette 1990), suricates (Clutton-Brock *et al.* 1999), primates (Treves 2000) and even mysid shrimp (Ritz 2000). Despite this, total group vigilance is generally higher in larger groups (Mooring *et al.* 2004, Quenette 1990) suggesting that individuals in smaller groups have a higher predation risk. Component Allee effects via this mechanism might be quite common in groups of mammalian herbivores, but to demonstrate them, individuals must be shown to pay a fitness cost of increased vigilance and/or to have higher predation mortality in small groups—both in principle likely but usually not proven. An exception is suricates (*Suricata suricatta*), which have strong, cooperative anti-predator vigilance. The combination of a high metabolism and small favoured prey (invertebrates) means that they need to forage for long periods of time. In presence of predators, each member of the group takes turn to stop foraging and assume the sentinel behaviour for the whole group. When the group becomes too small, however, the individuals can no longer stop foraging each time their turn comes, and long periods go by without sentinels, resulting in higher mortality rates in smaller groups. Adult survival increased with group size regardless of predator density, since these cooperative breeders cooperate in other ways as

Figure 2.13. In the suricate, *Suricata suricatta*, group vigilance is reduced in small groups, resulting in increased mortality. Photo: J. A. Barnard.

well (see below), suggesting another component Allee effect. These component Allee effects may lead to a demographic Allee effect, with extinction of groups smaller than nine in years of harsh conditions (low rainfall) (Clutton-Brock *et al.* 1999; Table 2.2).

Larger flocks of wood pigeons (*Columba palumbris*) detect birds of prey at a greater distance and individuals in larger flocks have higher survival as a consequence (Kenward 1978). Lapwings (*Vanellus vanellus*) are aggressive towards potential nest predators, a strategy which is more effective in large than small nesting aggregations—in fact both density and aggregation size seem to have independent positive effects on nest survival from predation by birds, although possibly not by foxes (Berg *et al.* 1992). Many seabirds such as gulls, terns and murres are likewise aggressive towards nest predators such as crows and gulls, contributing to a reduction in predation in the middle of dense colonies (Lack 1968, Gilchrist 1999).

Food web interactions: the cultivation effect

Predator–prey interactions occur in the context of a wider food web. This can be true even if (as here) we confine our discussion to one predator and one prey species, because both can play different roles in the food web as they progress through different ontogenetic stages. Piscivorous fish, for example, tend to feed rather indiscriminately on the basis of size rather than selecting distinct prey species. Generally, a smaller fish of any species is prey, a similar-sized fish is a competitor and a larger fish is a predator. This means that predator–prey interactions between two species can operate in both directions at once; Species B can interact with Species A simultaneously as a predator (of small juveniles), a competitor (of larger juveniles) and prey (of adults). A high density of Species A adults can thus protect Species A juveniles from predation by reducing the density of Species B.

This effect, sometimes termed the 'cultivation effect' (Wooten 1994, Walters and Kitchell 2001) or 'mutual predation' (Gardmark *et al.* 2003), is well known in lake fisheries where managers try to minimize the mortality of newly stocked juveniles (Walters and Kitchell 2001). It has also been suggested as an Allee effect mechanism in cod (see Box 2.5). Note, however, that many fish species are cannibalistic and thus prey on juveniles of their own species, which probably makes Allee effects generally less likely.

The idea of the cultivation effect can be extended to other types of passive protection afforded to juveniles by adults. Juveniles of the red sea urchin (*Strongylocentrotus franciscanus*) which settle from the plankton under adult spine canopies have a much higher survival than juveniles which settle in other benthic habitats (Tegner and Dayton 1977). This of course is more effective at high adult density.

Box 2.5.   **Allee effects in Atlantic cod?**

Overfishing has driven populations of Atlantic cod (*Gadus morhua*) to a few percent of historical abundances: the North Sea population has declined 90% since 1970, while populations around Newfoundland are at 1% of 1960s levels—a loss of at least 2 billion reproductive individuals from these areas alone (Rowe *et al.* 2004). The Newfoundland cod fishery was closed in 1992 with devastating social and economic consequences. Scientists were initially confident that populations would recover quickly after high fishing mortality was removed, since logistic population growth predicts high per capita population growth rates at low density (Roughgarden and Smith 1996). However, there is so far little evidence of an increase in population size since the moratorium, at least up to 2003 (Rowe *et al.* 2004), suggesting a low per capita population growth rate.

It is difficult in fish to measure reproductive output directly. Generally, abundance of juveniles of a certain size ('recruits') is used as a proxy, but of course, this cannot take account of the various demographic and environmental influences on juvenile numbers during early life history. A formal analysis of stock-recruit relationships found evidence of demographic Allee effects in only one out of 26 fish populations with sufficient data (Myers *et al.* 1995) although this may reflect limited statistical power and difficulties determining the geographical boundaries of a fish 'population' (Shelton and Healey 1999, Liermann and Hilborn 1997) (see Section 5.2).

Cod have quite complex reproductive behaviour. Females broadcast eggs into the water column where they are fertilized both by a primary male, who aligns his urogenital tract with hers, as well as by various satellite males. In lab experiments, the proportion of eggs fertilized increases with the number of satellite males present. In addition, the female and primary male need to be a similar size for egg and sperm ejaculation to occur close together and hence for fertilization by the primary male to be efficient. Thus if males are sparse and females have limited mate choice, fertilization will be less effective on two grounds (Rowe *et al.* 2004).

Cod may also have a component Allee effect in survival (Swain and Sinclair 2000, Walters and Kitchell 2001). There is a positive correlation between adult mortality (from fishing), juvenile mortality (Myers *et al.* 1997) and egg and larval mortality (Anderson and Rose 2001), consistent with the cultivation effect. There is also a negative correlation between the survival and recruitment of larval and juvenile cod and the biomass of pelagic fish such as herring and mackerel (Swain and Sinclair 2000), which feature fish

---

Box 2.5. (*Continued*)

---

eggs and larvae strongly in their diet, and in turn feature strongly in the diet of adult cod. Cod may even suffer from genetic Allee effects (see Section 4.1) and from emergent Allee effects if cod is the top predator (de Roos and Persson 2002; Section 3.6.1).

Allee effects in cod, however, remain conjectural, particularly since field observations of reproduction are difficult and the strength of food web links are largely unknown. Other phenomena probably also play a role in the recovery failure, including climate change (Rose 2004) and the generally destructive effects of bottom fishing on marine ecosystems (Jennings and Kaiser 1998, Hiddink *et al.* 2006).

Conclusions about prey aggregation

The costs and benefits of aggregation have caused controversy in ecology for many years (Møller 1987, Danchin and Wagner 1997). The fact that many species exhibit an aggregated distribution is a strong indication that there must be net benefits, particularly since individuals in large groups can bear significant costs associated with high density, including transmission of diseases and parasites (Brown and Brown 2002, 2004), competition for food, space or mates, extra-pair copulations (Brown and Brown 2003) and cannibalism or infanticide (Møller 1987).

When individuals are aggregated they normally face a trade-off between improved survival from predation (as well as other potential benefits such as increased mating and foraging opportunities) and reduced growth due to competition. Whether a component Allee effect in survival leads to a demographic Allee effect depends on the relative influence of these processes on the per capita population growth rate. This trade-off has been shown empirically in juvenile queen conch, which have higher growth rates but greatly decreased survival (due to predation) outside aggregations (Stoner and Ray 1993, Ray and Stoner 1994). The balance of costs and benefits comes down strongly in favour of aggregation in this species, suggesting a demographic Allee effect. Some species, including gerbils and schooling fish, actively judge this balance between predation risk and competition and alter their behaviour accordingly, separating to minimize competition when predation risk is low but clumping together if it is high (Rosenzweig *et al.* 1997, Rangeley and Kramer 1998).

2.3.3. **Aggregation by predators**

Predation-driven Allee effects can apply to parasitism as well as predation. A review of 171 insect host-parasite relationships showed an inverse relationship

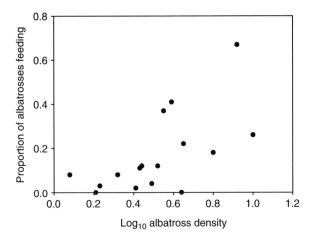

Figure 2.14.   The foraging success of black-browed albatrosses as a function of population density. Albatrosses at higher density spend more time feeding because they use the foraging success of conspecifics to find their highly patchy prey (krill swarms). Data from Grünbaum and Veit (2003); the positive relationship is statistically significant. (The nearest albatross nesting site is 1500 km away from where these foraging individuals were observed.)

between host population size and parasitization rate in 46 cases (Stiling 1987). A similar relationship was found between density of dickcissel (*Spiza americana*) and rates of nest parasitism by the cowbird (*Molothrus ater*) (Fretwell 1977). In other cases however, individuals at high density can suffer from higher levels of parasitism (Brown and Brown 2002, 2004).

We have discussed aggregation by prey species at length above, but predators can also gain fitness benefits from aggregation in hunting and foraging. As with prey there is a range of behavioural complexity. Predators may opportunistically use the presence of conspecifics as a signal that food is available, but may also have complex social and cooperative hunting behaviour (discussed separately below).

We have already mentioned improved foraging efficiency on highly patchy prey as a mechanism for coloniality in seabirds; it may also apply to other marine predators such as cetaceans, seals and sealions, which are also frequently social. A study on foraging by black-browed albatrosses (*Thalassarche melanophris*) on krill swarms (*Euphausia superba*) suggests that observing conspecifics foraging may be a more successful means of finding food than observing the prey directly (Fig. 2.14). Models suggest that this should result in a component Allee effect in fitness since foraging success will be higher at higher albatross density (Grünbaum and Veit 2003). A similar mechanism has been proposed to explain the spectacular collapse to extinction in 1914 of the passenger pigeon, since they relied on a highly patchy 'prey'—mast-fruiting trees. Colonial and possibly cooperative breeding may also have played a role (Halliday 1980).

2.4. **Allee effects in cooperative species**

Individuals of highly social species, by definition, have a need for conspecifics, whether for cooperative hunting and foraging, rearing of offspring, protection from predators, reproductive opportunities or some other mechanism, or (usually) a combination of many factors. It would seem logical to predict that increased sociality- i.e., increased need for conspecifics- would be correlated to increased susceptibility to Allee effects. Surprisingly, however, it is not in highly social species that Allee effects have been the most studied.

For example, many primate species are highly social, but the evidence for Allee effects in primates does not seem to be strong; they may arise in models but on closer inspection often turn out to be an assumption rather than a result (e.g. Swart *et al.* 1993). In fact, simple models suggest that Allee effects are more likely in solitary than social primate species (Dobson and Lyles 1989), seemingly contradicting our argument that Allee effects are more likely in gregarious species. These models are oversimplistic because they do not take evolution into account; solitary species are more likely to be at low density, but they are also adapted to low density. We return to this idea in Chapters 4 and 5.

Even within cooperative species, the evidence for Allee effects is quite small, even though logic points to the likely presence of component Allee effects, at the very least. A variety of species, mainly vertebrates but also some insects (most famously ants and bees) have evolved beyond aggregation to active cooperation with individuals playing different roles within the group. Generally, cooperation of this sort involves both survival (finding food) and reproduction simultaneously, which is why we have chosen to discuss it in a separate section. Some examples of cooperative species and the fitness benefits they gain from group size are shown in Table 2.3.

The lack of studies on Allee effects in cooperative species is probably the most noticeable in eusocial arthropods. It is common knowledge that ants use communication to locate and defend resources, that termites need large colony sizes to overcome those of ants, or that bees actively thermoregulate their hives with coordinated behaviours. Relatively little work is available, however, on the importance of colony size to the fitness of individuals and the growth of the colony. An exception is the work by Avilés and Tufiño (1998) on social spiders *Anelosimus eximius*, that shows that because offspring survival to maturity increases with colony size, the lifetime reproductive success of female social spiders decreases in small colonies. Indirect evidence for Allee effects in eusocial insects is considered in Box 2.6.

Similarly, most cooperative breeders, where not all adults reproduce and subordinate individuals act instead as 'helpers', have not benefited from studies aiming at demonstrating or refuting an Allee effect, with the notable exceptions of

Table 2.3. Fitness benefits to large group size in cooperative species (adapted from Courchamp *et al.* 2000b; for references see original paper).

Fitness benefit	Cooperative process	Examples
Improved survival	More efficient at finding food	Varied sittella, evening bats
	More efficient at catching prey	Harris hawks, lions, spotted hyena, African wild dogs
	Feeding and defence of sick or wounded individuals, wound-licking	African wild dogs
	Communal anti-predator guarding	Florida scrub jays, striped-back wrens, dwarf mongooses, suricates
	Communal mobbing and defence of attacked individuals	White-winged choughs, Florida scrub jays, banded mongoose, Geoffroys marmoset
Increased reproduction through reduced parental effort	Increased reproductive life-span of breeders	Chimpanzees, pied kingfisher, splendid fairy wren, dwarf mongoose
	Shorter intervals between breeding events	Splendid fairy wren, prairie voles, house mice
	Increased litter size	Black-backed jackals
Increased offspring survival	Offspring feeding	White-fronted bee eaters, Arabian babblers, meerkats
	Offspring defence	African wild dogs
	Den or nest guarding, baby-sitting	African wild dogs, Florida scrub jay, bay-winged cowbird, coyotes, suricates
	Offspring grooming and warming	Red-billed wood hoopoes, pine voles
	Increased breeding experience	Splendid fairy wren, Seychelles warbler, several primate species
Various	Communal mobbing or coercion of competitors, kleptoparasites and brood parasites	Banded mongoose, spotted hyenas, splendid fairy wrens, acorn woodpeckers, African wild dogs

---

Box 2.6. **Allee effects in ants?**

The first evolutionary step for social insects on the road to becoming formidable super-organisms was to aggregate to escape unfavourable biotic and abiotic conditions, as in the many examples given in Section 2.3.1 above. It is known that in these species colony size is a critical variable, and this is particularly true for new colonies, which often need a minimum number of workers in order to have a reasonable chance of enduring. It is thus

Box 2.6. *(Continued)*

surprising that Allee effects have been so neglected in this field, particularly given the impressive community of researchers working on other aspects of eusociality in insects. Specialists in bees, wasps, termites and ants seem to agree on the fact that logistical and conceptual obstacles specific to these species are a real disadvantage for such studies. However, some data can be found that suggest that Allee effects should certainly be worth studying in social insects (as well as in other eusocial organisms). We develop these points briefly below taking ants as an illustration.

In most ant species there is a colony size threshold for reproduction (a minimum number of workers is needed to attain the reproductive stage for sexuate production, Hölldobler and Wilson 1990). Colony fitness (measured as sexuate production) is frequently linked to colony size (Hölldobler and Wilson 1990, Tschinkel 1993, Gordon 1995). In some cases, colony size significantly affects the probability of reproduction but not the amount of reproduction (Cole and Wiernasz 2000). Many species show a positive correlation between colony size and measures of colony reproduction (*Tetramorium caespitum*, Brian and Elmes 1974; *Myrmica sabuleti*, Elmes and Wardlaw 1982; *M. sulcinodis*, Elmes 1987; *Harpagoxenus sublaevis*, Bourke *et al.* 1988; *Lasius niger*, for some measures, Boomsma *et al.* 1982; *Trachymyrmex septentrionalis*, Beshers and Traniello 1994; *Pheidole 'multispina,' P. 'rugiceps,'* Kaspari and Byrne 1995; *Formica podzolica*, Savolainen *et al.* 1996), however this does not exclude the possibility that the onset of reproduction is geared to colony age. Colonies of legionnaire army ants function under quite different principles on many grounds (Hölldobler and Wilson 1990), but they too rely on numbers, notably for foraging, so that Allee effects could be determinant at the onset of colony growth in these species, too.

The probability of colony survival too is related to the colony size (ant biomass). In many cases the size of the nest is inversely correlated with the probability of death (Wiernasz and Cole 1995, Adams and Tschinkel 2001). Established colonies (older and greater) suppress the establishment of new colonies (after queen arrival); in many ant species there exists empirical evidence that older colonies affect new colonies establishment (Ryti and Case 1988, Wiernasz and Cole 1995, Gordon and Kulig 1996, MacMahon *et al.* 2000; Billick *et al.* 2001). However, competition with neighbours rarely causes the death of established colonies: once a colony is established, its neighbours have little effect on its survival (Gordon and Kulig 1998).

Figure 2.15. The African wild dog, *Lycaon pictus*, an endangered obligate cooperative breeder for which several component Allee effects have been demonstrated. Photo: Gregory Rasmussen.

suricates and African wild dogs. The importance of the number of 'helpers' has been shown in a number of species, such as white-winged chough (*Corcorax melanoramphos*, Boland *et al.* 1997), splendid fairy-wrens (*Malurus splendens melanotus*, Russell and Rowley 1988), white-fronted bee-eaters (*Merops bullockoides*, Emlen and Wrege 1991), Arabian babblers (*Turdoides Squamiceps*, Wright 1998), Florida scrub jays (*Aphelocoma coerulescens*, MacGowan and Woolfenden 1989), acorn woodpecker (*Melanerpes formicivorus*, Mumme and DeQueiroz 1985), dwarf mongooses (*Helogale parvula*, Waser *et al.* 1995) and Damaraland mole-rats (*Cryptomys damarensis*, Jarvis *et al.* 1998), to name a few. Globally, these studies suggest that where helpers are important, individuals in small groups suffer from reduced fitness.

In addition to the role of helpers in raising young, the best examples of putative Allee effects in cooperative species probably come from cooperative hunting in social carnivores such as hyena (*Crocuta* spp.), lions (*Panthera leo*), wolves (*Canis lupus*), coyote (*C. latrans*), black-backed jackals (*C. mesomelas*), dhole (*Cuon alpinus*) and African wild dog. Cooperative hunting in these species is probably related to two main factors: the difficulty of capturing large or dangerous prey and competition with scavengers and kleptoparasites. Suricates forage alone for small invertebrates but cooperate to kill dangerous snakes. Small packs of African wild dog frequently lose their kill to spotted hyenas (Gorman *et al.* 1998), who in turn often lose their kill to lions when their groups are small and lions abundant (Trinkel and Kastberger 2005). Models and data suggest that in the absence of scavengers, wolves would gain most energetically by foraging alone or in pairs or small groups, but the need to defend kills from ravens (who can consume up to half of a kill) means that the optimal pack size increases significantly (Thurber and Peterson 1993, Vucetich *et al.* 2004).

Obligate cooperative breeding has more complex evolutionary origins, related to limited reproductive options (e.g. lack of territories) and benefits accruing to helpers either via improving their later reproductive success (and lifetime reproductive fitness) and/or via kin selection (Emlen 1997; Table 2.3). However, molecular data suggests that individuals in spotted hyena clans cooperate in some ways even when not particularly closely related, so that fitness benefits of group living do not necessarily derive entirely from altruism towards kin (Van Horn *et al.* 2004). In other species, such as African wild dog, pack members are closely related and group selection has clearly played an important role (Girman *et al.* 1997) although adoption of unrelated pups by small groups has been repeatedly observed (Courchamp and Macdonald 2001).

The role of cooperative reproduction in relation to group size and Allee effects has been studied most carefully in the African wild dog, spurred by a realization that many of the remaining populations were decreasing in size. African wild dogs have a strongly developed system of cooperative breeding, where a single breeding pair in each pack raise their young with help from other adult relatives in hunting, pup feeding and pup guarding; single pairs are usually unsuccessful at raising pups. Courchamp and Macdonald (2001) provide a list of multiple benefits of large pack size in the African wild dog, via both cooperative hunting (more efficient hunting, less risk of injury, defence from kleptoparasites), cooperative breeding (larger litters, better pup care, higher pup survival), general benefits of group living (anti-predator vigilance, competition for resources with other packs) and altruism towards kin (tolerance of wounded and old animals at kills, wound licking).

Small groups (five adults or fewer) suffer from a trade-off between these various roles in cooperative hunting and breeding. If an adult is left as a pup guard ('babysitter') during hunting trips, the hunt is less likely to be successful and the kill more likely to be lost to kleptoparasites; as a consequence they face nutritional deficit. If no babysitter is left, pups will suffer high mortality, again often from kleptoparasite species such as hyena and lions (Courchamp *et al.* 2002). These small packs do not seem to be viable. This demographic Allee effect at the pack level (Angulo *et al.* 2008) has knock-on effects at the metapopulation level, since larger packs supply larger dispersing cohorts, which are more likely to be successful in establishing new packs; a metapopulation model suggests that this gives rise to an 'Allee threshold' in the number of packs, below which the metapopulation will go extinct because new packs cannot be formed fast enough (Courchamp *et al.* 2000a). They suggest that Allee effects may be likely in species which have a strong social structure, and an analysis of extinction rates in carnivores since the Miocene suggests that highly social taxa are more likely to go extinct than taxa which are solitary or live in small family groups (Munoz-Duran 2002).

2.5. **Conclusions**

There are many potential mechanisms for Allee effects, most of which are discussed above (some we may have missed). These mechanisms (component Allee effects) are themselves of interest to ecologists, particularly from the evolutionary perspective; have species evolved to counteract these component Allee effects and if so how? They also interact in complex ways with other ecological processes such as dispersal, and are further complicated by issues such as population and landscape spatial structure and spatial scale (Box 2.7). These issues are considered further in the remaining chapters. From a more applied point of view, the key question is whether a mechanism which gives rise to a component Allee effect can in turn lead to a demographic Allee effect and thus to the potential for extinction thresholds.

Some mechanisms operate only under certain conditions; obviously predation-driven Allee effects only operate in the presence of the predator, for example. Other mechanisms may operate in all populations, but we have seen that the benefits of high density may trade off with costs. Plants experiencing high pollination rates may exhaust their resources to set seed, sea urchins at high density have high fertilization efficiency but are smaller and produce fewer gametes, and juvenile conch trade slower growth for survival benefits in aggregations. There may be benefits as well as costs to low density even if an Allee mechanism is present—in the island fox, low density results in low pup survival but more pups per female (Fig. 2.6). In general, different components of fitness are likely to have different relationships with population size or density, and thus the link between component and demographic Allee effects is, as we stress, by no means direct.

You may by now have noticed some overlap in the mechanisms we used as examples. Some mechanisms act in several different ways simultaneously. Coloniality in birds, for example, has benefits for predator dilution and for finding patchy resources and (for females) in offspring quality through the high frequency of extra-pair copulations in some cases (Danchin and Wagner 1997). Synchronous flowering reduces both pollen limitation and seed predation. Does a pregnant female African wild dog waiting at the den for the return of the hunting party to feed her gain a fitness benefit in survival or in reproduction? A single mechanism can have multiple fitness benefits. Some species also have more than one mechanism operating at the same time, providing the basis for a complicated web of costs and benefits at low density (Berec *et al.* 2007). This is not surprising, since once there is some fitness benefit to living in groups, and groups start

Box 2.7.   **Spatial scale and density dependence**

The relationship between predation rates and prey density or population size depends to a significant extent on the spatial scale of measurement. For example, predator functional response can vary according to the spatial scale of measurement (Morgan *et al.* 1997). Likewise, density-dependent relationships between predators and prey can look different at different spatial scales. A study of cicada attacks on dogwood trees (Cook *et al.* 2001), for example, showed that at a small scale (within a given woodland area), a higher proportion of dogwoods were attacked as dogwood density decreased (a component Allee effect in the trees via predator dilution). However, at a larger scale, a lower proportion of dogwoods were attacked in small areas of woodland relative to large area (negative density dependence in the trees via predator aggregation). This is because the predators (cicadas) were mobile enough to aggregate strongly in large patches of suitable 'prey' (i.e. there was a trade-off between cicada functional response at the small scale and cicada aggregative response at the larger scale). This leads to a component Allee effect at the tree (sub)population (woodland) level, but negative density dependence at the tree metapopulation (landscape) level.

This issue is not specific to predation, but arises in all Allee effect mechanisms. For instance, in his study of pollination, Lennartsson (2002) found that pollination rates increased with plant patch size, and also with the density of patches in the landscape. In this case, density dependence had the same sign at both scales, suggesting that the Allee effect in this case is reinforced by mechanisms working at different spatial scales; not only patch size but also patch isolation and the size of the whole habitat area are important. Population size, density and isolation have also been shown to interact in this way in studies of *Clarkia concinna*, where isolation only reduces seed set in small populations (Groom 1998) and in sea urchins, where density can be critical to fertilization efficiency in small populations but not large ones (Levitan *et al.* 1992, Levitan and Young 1995).

The spatial scale at which we measure population size or density can thus determine the nature of density dependence and the sign of interactions with other individuals. Density dependent mechanisms at different spatial scales may reinforce or counteract each other.

Table 2.4. Some potential examples of species with more than one Allee effect mechanism (adapted from Berec *et al.* 2007)

Species	Multiple component Allee effects	References
Marsh gentian *Gentiana pneumonanthe*	Small populations have lower fecundity owing to a reduced pollination success and increased inbreeding	Oostermeijer (2000)
Monarch butterfly *Danaus plexippus*	Individuals in small overwintering groups suffer multiple costs: mating depression in spring, less efficient predator dilution, and decreased protection against cold	Calvert *et al.* (1979) Wells *et al.* (1998)
Social spider *Anelosimus eximius*	Small colonies have a reduced number of eggs per sac[a] and reduced offspring survival to maturation (owing to reduced protection against predators and low ability of colonies to acquire resources)	Avilés and Tufiño (1998)
Red sea urchin *Strongylocentrotus franciscanus*	Individuals in low-density populations have reduced fertilisation efficiency and survival of new recruits owing to a cultivation effect[b] (adult urchins, when at high densities, protect juveniles from predation in their canopy of spines and facilitate their feeding)	Levitan *et al.* (1992), Tegner and Dayton (1977)
Ribbed mussel *Geukensia demissa*	Low-density populations suffer from increased mortality owing to crab predation and winter ice	Bertness and Grosholz (1985)
Atlantic cod *Gadus morhua*	Individuals in small-sized populations have reduced fertilization efficiency and juvenile survival owing to a cultivation effect[b] (adult cod prey on species that prey on juvenile cod; fewer adult cod imply higher juvenile mortality)	Rowe *et al.* (2004), Walters and Kitchell (2001)
White abalone *Haliotis sorenseni*	Individuals at small density have reduced fertilization efficiency; some populations suffer overexploitation and the species being considered a luxury item is subject to an anthropogenic Allee effect	Courchamp *et al.* (2006), Hobday *et al.* (2001), Babcock and Keesing (1999)
Lesser kestrel *Falco naumanni*	Small colonies suffer from increased nest predation and lower adult survival[c]	Serrano *et al.* (2005)
Speckled warbler *Chthonicola sagittata*	Small overwintering flocks might face reduced vigilance to predators[b] and foraging efficiency[b]	Gardner (2004)
Sooty shearwater *Puffinus griseus*; Hutton's shearwater *P. huttoni*[b]	Birds in small colonies have lower breeding success and higher mortality[c], both owing to predation by the stoat *Mustela erminea*	Cuthbert (2002)
African wild dog *Lycaon pictus*	Smaller packs are less efficient in cooperative breeding, cooperative anti-predator behaviour, and cooperative hunting	Courchamp and Macdonald (2001)
Southern fur seal *Arctocephalus australis*	Small populations face reduced pup survival owing to habitat change (plain to more rugged) elicited by human disturbance and to predation by southern sea lions *Otaria flavescens*; these effects were not observed in the same population	Stevens and Boness (2003), Harcourt (1992)

Table 2.4. *(Continued)*

Species	Multiple component Allee effects	References
Southern sea lion *Otaria flavescens*	Small breeding groups suffer a 'constant-yield-like exploitation' by killer whales *Orcinus orca*[b] and a reduced survival of pups caused by displaced mating behaviour of adult males[b,c]	Thompson *et al.* (2005)
Alpine marmot *Marmota marmota*	Individuals in smaller overwintering groups have difficulties in finding mates and decreased survival owing to less efficient social thermoregulation	Stephens *et al.* (2002a)
Desert bighorn sheep *Ovis canadensis*	Individuals in smaller groups suffer higher predation risk because of lower vigilance and the species rarity makes it a valuable item for trophy hunters	Courchamp *et al.* (2006), Mooring *et al.* (2004)

[a] Exact mechanism unknown.

[b] These mechanisms have been hypothesized by the authors of the respective studies.

[c] The same mechanism operates on two distinct life stages.

to form, natural selection can then act to increase the fitness advantages of living in groups by selecting for novel forms of cooperation between group members.

The population dynamics consequences of these multiple mechanisms can be complex, and are discussed in detail in Section 3.2.2, but it is reasonable to assume (in general) that these species will have a higher probability of a demographic Allee effect than species with only one mechanism. Examples are given in Table 2.4.

Mechanisms for component Allee effects are many and varied. However, so are mechanisms for negative density dependence (e.g. competition for food, mates, space or territories, mutual aggression, direct physical interference...). A component Allee effect is only one of several density-dependent mechanisms acting on a population. The net effect of population size or density on overall fitness is thus not easy to predict, even if an Allee mechanism is present. Thus it is hard to determine just from the presence of a mechanism that a demographic Allee effect exists. In Chapter 3 we consider how modelling can be used to draw demographic conclusions from field measurements of component Allee effects, in Chapter 5 we look at various pragmatic means of assessing the probability of component and demographic Allee effects in a population and we consider the associated difficulties in Chapter 6. Nonetheless, a mechanism has to exist for an Allee effect to exist, so a study of component Allee effects is the key first step in all these processes.

3. *Population dynamics: modelling demographic Allee effects*

Much of what we know about Allee effects is from mathematical models. This is not surprising since models help us organize, conceptualize and interpret a vast amount of complex ecological data, and predict or hypothesize where such data are not available. They support our understanding of many within- and between-population processes, and supply us with well-founded, testable predictions regarding their dynamics. Applied ecologists often benefit from modelling in comparing impacts of alternative management scenarios on threatened or exploited populations that are not readily accessible to experimentation. Owing to the difficulties in identifying demographic Allee effects in the field (populations rarely hang around in the zone of reduced growth rate—either getting lucky and growing out of that zone, or unlucky and going extinct), models have been an important means by which to assess the potential importance of Allee effects for population and community dynamics.

This chapter deals with models of demographic Allee effects. We are going to show what types of model are available, what their underlying assumptions are, how they can be used—and most importantly, how they have contributed to our understanding of the dynamical consequences of Allee effects, both in single-species populations and multiple-species communities. As we aim this chapter to be accessible to a general readership, we conceived it as a sort of summary of a large amount of literature that has been published on models which address Allee effects: a comprehensive (if a little dense) overview of the topic. We do not provide detailed information on how to develop and analyse any single model, nor discuss the structure and predictions of every model that has ever been used to study Allee effects. Also, this chapter does not consider models of evolution with respect to Allee effects, an issue that is touched upon in Chapter 4.

To help a less mathematically oriented reader to keep pace with the text, some of the modelling jargon used in this chapter is given a short informal definition in Box 3.1. For more on principles of modelling population dynamics, the reader

---

Box 3.1. **Glossary of some of the modelling jargon used in this chapter**

Chaotic oscillations: Population trajectories of deterministic models that fluctuate, never attain an equilibrium and never visit any population density (size, state distribution) again; two trajectories starting close to one another may deviate considerably after sufficient time has elapsed.

Continuous-time model: Population model in which target species is assumed to reproduce continuously in time; analytical models defined in terms of differential equations.

Deterministic model: Population model in which any current population state determines a unique trajectory; population model free of any stochastic (random) element.

Discrete-time model: Population model in which reproduction of target species is pulsed (usually once a year in a relatively short time interval); analytical models defined in terms of difference equations; time runs in discrete steps (usually one year or generation).

Equilibrium: Population density (size, state distribution) that, once attained, does not further change in time.

Locally stable equilibrium: Equilibrium that attracts all population trajectories starting at least in its immediate vicinity: when slightly moved away from this equilibrium, the population comes back to it.

Population model (also model of population dynamics): Formal (mathematical) description of how population density, size and/or state distribution evolve in time, either in the form of dynamical equations (analytical models) or behavioural and demographic rules (simulation models).

Population stability: The concept of population stability can be viewed from various perspectives. When we say that population stability is reduced, we may mean that the population equilibrium changes from (locally) stable to unstable or that it takes longer to reach equilibrium. It may also mean that the equilibrium population density has been reduced, so that a smaller perturbation away from this equilibrium is now sufficient for the population to go extinct, or that the amplitude of population fluctuations has increased. Last but not least, we may say that population stability decreases once the range of model parameters for which the population goes extinct gets larger.

Box 3.1.   (*Continued*)

Simulation models: Mostly population models which treat individuals as discrete entities and where population dynamics are no longer defined in terms of equations, but emerge from a set of behavioural and demographic rules repeatedly applied to each individual.

State-structured population model: Model which incorporates population structure by classifying population members as belonging to a finite set of states; these states may include age, sex, developmental stage, body size, local density or any combination of these.

Stochastic model: Population model comprising a random element; no two trajectories starting from an equal initial condition are the same.

Strategic model: Model used to address general ecological questions such as effects of space, gender or predation.

Sustained oscillations: Population trajectories of deterministic models that are not in an equilibrium but that go through some population densities (sizes, state distributions) over and over again.

Tactic model: Model used to understand or predict dynamics of a specific population with sufficient precision. Often needs to be more realistic which also makes it more complex.

Unstable equilibrium: Equilibrium that repels at least one population trajectory starting some distance away from the equilibrium, regardless of how small that distance is: even when slightly moved away from this equilibrium, the population does not come back to it.

may consult many recent books on theoretical ecology such as Gotelli (1998), Case (2000), Kot (2001) or Turchin (2003).

Models play four essential roles in the study of Allee effects. They are used to:

- formally describe (empirical observations of) component and demographic Allee effects in the form of an equation,

- assist us in searching for component and demographic Allee effects in empirical data,

- assess whether component Allee effects give rise to demographic Allee effects,

- predict wider consequences of demographic Allee effects for population and community dynamics.

To play these roles properly, models enter the realm of Allee effects in several distinctive forms—phenomenological models of demographic Allee effects, component Allee effect models, and structured (in various ways) population models. Sketching their conceptual importance here, and outlining how they combine to produce predictions on demographic Allee effects and their implications for population and community dynamics, we hope to give the reader a clear vision of what lies ahead in this chapter.

We start with deterministic models of demographic Allee effects. Demographic Allee effects can be modelled in a number of different ways. Firstly, there are phenomenological models which tend to describe the per capita population growth rate by an *ad hoc*, hump-shaped function of population size or density (Fig. 1.9). The essence of this approach goes back to Odum and Allee (1954) who appear to have been the first to convey the idea of a hump-shaped per capita population growth rate in the form of an equation. Some of the simpler of these models assist us in searching for demographic Allee effects in census data on population size or density (e.g. Liebhold and Bascompte 2003).

Secondly, we have more mechanistic models which arise by inserting models of component Allee effects into suitably structured population models (state-structured or simulation models) and help us assess various demographic consequences of component Allee effects, including their potential to generate demographic Allee effects. These kinds of models were used to explore, e.g., how poor conditioning of the environment in low-density populations of the fruit fly *Drosophila melanogaster* affected their long-term dynamics (Etienne *et al.* 2002) or the impact of pollen limitation in low-density patches of the smooth cordgrass *Spartina alterniflora* on its invasion rate (Taylor *et al.* 2004). Component Allee effect models, by themselves, serve also to formalize evidence on diverse component Allee effects, as exemplified by Dennis (1989) or Tcheslavskaia *et al.* (2002) on mating efficiency as a function of population density.

Next, we take into account the fact that deterministic models of demographic Allee effects are not always appropriate, given that Allee effects often affect small populations where stochasticity cannot usually be disregarded. Most importantly, stochastic models of strong demographic Allee effects, unlike deterministic models, predict that any population which drops below the Allee threshold may still grow and persist, while any population which starts above it may eventually go extinct (e.g. Allen *et al.* 2005).

Finally, scaling up one level, we consider the role of demographic Allee effect models as building blocks for other structured population models, namely space-structured population models and multiple-species community models. In this way, for example, Amarasekare (1998b) studied how dispersal between two local populations affected survival of the metapopulation as a whole and Courchamp *et al.* (2000b) assessed the impacts of natural enemies on a species with a strong demographic Allee effect.

On the tour of Allee effect models, we realize that in some (structured) population models, demographic Allee effects can emerge even without treating an underlying component Allee effect explicitly, i.e. without using any component Allee effect model. Such demographic Allee effects are sometimes referred to as *emergent* Allee effects. Models showing emergent Allee effects are still rare, but include some models of single-species populations (e.g. simulation models of sexual reproduction; Berec *et al.* 2001) and models of multiple-species communities (e.g. predator–prey models with stage-selective predation; de Roos *et al.* 2003). We emphasize the occurrence of emergent Allee effects where appropriate.

As the reader will discover when reading this chapter, numerous models have been used to study Allee effects, and others will undoubtedly be developed. The choice of a specific model (including the depth of model detail) depends crucially on the question posed and available knowledge on the modelled population(s). Simplicity is preferred when the model is to be strategic, that is, used to address general ecological questions such as effects of space, gender or predation. Tactic models, i.e. models used to understand and/or predict dynamics of a specific population with sufficient precision, need to be more realistic, which requires them to be more complex. In general, modelling is a subjective procedure driven by an objective goal: through an appropriate model to understand past and predict future dynamics of populations and communities.

3.1. Phenomenological models of demographic Allee effects

Among the initial choices modellers have to make when modelling population dynamics is how to handle time. There are two mathematical frameworks which correspond to a perception of time either as continuous—ordinary differential equations—or as discrete steps (often one year or generation)—difference equations. We speak of continuous-time and discrete-time models, respectively. In both cases, dynamics of a population is captured via a function $g(N,X)$ which represents the per capita population growth rate; N is population size or density and X represents all the other things affecting population growth, such as food, natural enemies, weather etc. Assuming, as usual, that X remains constant over time, most continuous-time models of population dynamics have the form

$$\frac{dN}{dt} = Ng(N) \tag{3.1}$$

while discrete-time models look like

$$N_{t+1} = N_t g(N_t) \tag{3.2}$$

In both equations, t represents time. When a population suffers a demographic Allee effect, $g(N)$ has a hump-shaped form (Fig. 1.9). A wide range of functions has been used in the literature to describe such a hump-shaped form; a representative selection of these functions, together with their basic properties, is set out in Table 3.1. As these functions have been driven more by mathematical tractability than by testing their match with real data or by derivation from biological processes, we refer to the resulting models as phenomenological models of demographic Allee effects.

Is there any preferred form of $g(N)$? Not an easy question. Functions differ greatly in their flexibility and ability to describe demographic Allee effects. Some general guidelines can nevertheless be outlined. If one's study aims to examine the impact of Allee effect presence versus absence in a population, a good model should admit pure negative density dependence as a special case. In other cases, one may prefer to cover both weak and strong Allee effects in a single model, so as to allow for a cogent comparative study of impacts of these types of Allee effect. Finally, one should consider allowing the Allee threshold and the carrying capacity of the environment to be explicitly included as model parameters if an assessment of demographic Allee effects of different strength is a priority. As an initial choice, we propose to consider the continuous-time model (P4) of Table 3.1,

$$\frac{dN}{dt} = rN\left(1 - \frac{N}{K}\right)\left(1 - \frac{A+C}{N+C}\right) \tag{3.3}$$

or its discrete-time counterpart (P8), both satisfying all the above 'requirements'. In model (3.3), r denotes the maximum per capita population growth rate in absence of the Allee effect, A is the Allee threshold, and K is the carrying capacity of the environment. The remaining parameter C affects the overall shape of $g(N)$—as C increases $g(N)$ becomes increasingly 'flatter' and reaches lower maximum values (Fig. 3.1). The scenario in which the demographic Allee effect is absent corresponds to $A = -C$.

The highest flexibility among the models listed in Table 3.1 is provided by model (P7) developed by Jacobs (1984). This flexibility is, however, gained at the cost of having six parameters that need to be estimated in empirical studies and tested for sensitivity in theoretical studies. On the other hand, highly flexible models might be necessary for understanding transient (as opposed to sustained) dynamics as they depend on finer growth rate curve characteristics, and might

Table 3.1. Phenomenological models of demographic Allee effects.

Per capita growth rate $g(N)$	Parameter constraints	Types of Allee effect [conditions]	Selected references	Code
Continuous-time models (of continuously reproducing populations): $dN/dt = N\,g(N)$				
$r-a(N-b)^2$	$a,b > 0$	Fatal[†], Strong, Weak	Edelstein-Keshet (1988)	P1
$r(1-N/K)(N/K-A/K)$	$r,K > 0$	Strong [$A>0$], Weak [$A\leq0$]	Lewis and Kareiva (1993), Amarasekare (1998a, 1998b), Keitt et al. (2001), Morozov et al. (2004)	P2
$r(1-N/K)(N/A-1)$	$r,K,A > 0$	Strong	Gruntfest et al. (1997), Courchamp et al. (1999a)	P3
$r(1-N/K)(1-(A+C)/(N+C))$	$r,K,C > 0$	Strong [$A>0$], Weak [$A\leq0$, $C>|A|$]*	Wilson and Bossert (1971), Courchamp et al. (1999b, 2000a,b), Brassil (2001), all for $C = 0$, Boukal et al. (2007)	P4
$r(1-(N/K)^Q)(1-((A+C)/(N+C))^P)$	$r,K,C,P,Q > 0$	Strong [$A>0$], Weak [$A\leq0$, $C>|A|$]*	Berryman (2003), for $C = 0$	P5
$b+(a-N)/(1+cN)N$	$a,c > 0$	Fatal/Strong [$b<0$]*, Weak [$b\geq0$]	Takeuchi (1996)	P6
$[r+uv(N^w/(N^w+v))] - cN^z$	$u,v,w,c,z > 0$	Fatal, Strong, Weak	Jacobs (1984)	P7
*Discrete-time models** (of populations with pulsed reproduction): $N_{t+1} = N_t\,g(N_t)$*				
$e^{[g(N)]}$, where $g(N)$ takes any of the forms suggested for continuous-time models	equivalent to the respective continuous-time form	equivalent to the respective continuous-time form	Liebhold and Bascompte (2003), Tobin et al. (2007), both with the form (P2)	P8
$Gb/((N-T)^2+b)$	$b,T,G > 0$	Fatal [$G<1$]*, Strong/Weak [$G>1$]	Asmussen (1979)	P9
$\rho N/(a+N^2)$***	$\rho,a > 0$	Fatal [$\rho<2\sqrt{a}$], Strong [$\rho>2\sqrt{a}$]	Hoppensteadt (1982)	P10

$N^\gamma e^{(r-cN)}$	$c > 0, 0 < \gamma \leq 1$, any r	Fatal $[e^r < (ec/\gamma)^\gamma]$, Strong $[e^r > (ec/\gamma)^\gamma]$	Asmussen (1979), Avilés (1999)	P11
$N^\gamma r(1-N/K)$	$0 < r < 4, K > 0, 0 < \gamma \leq 1$	Fatal $[r < (1+\gamma)^{1+\gamma}/(K\gamma)^\gamma]$, Strong $[r > (1+\gamma)^{1+\gamma}/(K\gamma)^\gamma]$	Avilés (1999)	P12
$N^\gamma + r(1-N/K)$	$r, K > 0, 0 < \gamma < 1$	Fatal/Strong $[r<1]$, Weak $[r>1]$	Avilés (1999)	P13
$a(N) f(N)$ – a general class of models	$a(0) \geq 0, da/dN > 0, \lim_{N\to\infty} a(N) = 1$ $f(0) > 0, df/dN < 0, 0 \leq \lim_{N\to\infty} f(N) < 1$		Burgman et al. (1993), Scheuring (1999), Fowler and Ruxton (2002), Schreiber (2003)	P14

Conditions for parameters in square brackets are given only in simple cases. These conditions highlight the parameters or parameter combinations which generate demographic Allee effects. Parameters in the phenomenological models generally do not allow for any specific biological interpretation. Adapted from Boukal and Berec (2002).

† By fatal Allee effects we understand the case in which $g(N)$ is hump-shaped but as a whole lies below the density axis—$g(N)<0$ for all $N \geq 0$ no matter how large or dense the population is, it declines and eventually goes extinct.

* Instances not considered in the original papers.

** In discrete-time models, the equilibrium N_K corresponding to the carrying capacity of the environment loses its stability, giving rise to sustained or chaotic oscillations, provided that $|1+N \, dg/dN|$ evaluated at $N = N_K$ is larger than 1 (e.g. Case 2000).

*** This function can be generalised to $\rho N^{d-1}/(a+N^d)$, $d \geq 1$ ($d = 1$ means no Allee effect), the form used in fisheries to model stock-recruitment relationships and known as the modified Beverton-Holt model in that field (e.g. Myers et al. 1995). In this latter form it is more a description of a component Allee effect affecting reproduction and/or early survival, and is therefore discussed in Section 3.2.1.

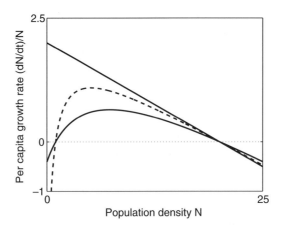

Figure 3.1. Per capita population growth rate as a function of population density for model (3.3) and different parameterizations of strong Allee effect. Straight line corresponds to no Allee effect ($A = -C$, $C = 5$), curved solid line ($A = 1$, $C = 5$) and dashed line ($A = 1$, $C = 0.5$) represent two strong Allee effects. Note that although location of the Allee threshold is the same in both cases, as C increases the curve becomes increasingly 'flatter' and reaches lower maximum value, and the population density at which the per capita growth rate is maximized moves to a higher density.

play a role especially in space-structured and multiple-species models. The ideal characteristics to tune a hump-shaped function $g(N)$ would be N_A (Allee threshold), N_K (carrying capacity), N_{max} (optimum size or density at which $g(N)$ is maximized), $g(0)$, $g(N_{max})$, and possibly also the slope of $g(N)$ at N_A and N_K. In fact, available models are limited only by imagination. However, many adequate forms have already been explored and their basic dynamics are reasonably well understood, so that we might as well stick with those unless a need arises.

3.2. From component Allee effects to demographic Allee effects

3.2.1. Component Allee effect models

Component Allee effect models relate a fitness component of individuals in a population to population size or density in a quantitative way. The majority of available component Allee effect models quantify either mate-finding Allee effects or predation-driven Allee effects. Although many of their underlying assumptions are unlikely to hold literally in nature, these models may prove sufficiently flexible to fit a variety of actual observations (e.g. Dennis 1989, Pfister and Bradbury 1996, Tcheslavskaia *et al.* 2002). We discuss these two classes of component Allee effect models in more detail below, together with stock-recruitment models of Allee effects in reproduction and/or early survival commonly used in fisheries, and mention other models only briefly. Several other reviews of component Allee effect models are available (Dennis 1989, Liermann and Hilborn 2001, Boukal and Berec 2002, Gascoigne and Lipcius 2004a), which an interested reader may consult for more details.

Models of mate-finding Allee effects

The component Allee effect arising as a result of a need to find mates is in a sense the flagship of component Allee effects—there is hardly any work on Allee effects that would not mention mate finding among the main Allee effect mechanisms. Almost invariably, modelling the mate-finding Allee effect involves some function $P(M,F)$, interpreted in the continuous-time framework as the female mating rate and in the discrete-time framework as the probability that a female mates within a given time step; M and F denote male and female population density, respectively. To represent the mate-finding process appropriately, $P(M,F)$ should satisfy a few generic properties:

• there is no mating if there are no males—$P(0,F) = 0$ for any F

• for a given number of females, a female's probability of mating (or mating rate) cannot decrease if the number of males increases—$P(M,F)$ is a non-decreasing function of M for any fixed F

• for a given number of males, a female's probability of mating (or mating rate) cannot increase if the number of females increases—$P(M,F)$ is a non-increasing function of F for any fixed M

• if males greatly outnumber females mating is virtually certain—$P(M,F)$ approaches 1 for a sufficiently large M/F ratio

The null case of no mate-finding Allee effect corresponds here to setting $P(M,F) = 1$ for any positive M and F, i.e. to the assumption that even one male suffices to fertilize all females.

The popularity of the mate-finding Allee effect has resulted in a large number of specific $P(M,F)$ forms. Table 3.2 provides a list of the simplest ones, together with their underlying assumptions. Functions (C1) and (C2) of Table 3.2 have been most frequently used in strategic models, presumably due to their simplicity and plausibility of underlying assumptions. It is relatively easy to show that (C2) follows from (C1) if the assumption of an equal search rate for all population members in (C1) is replaced by that of an exponentially distributed search rate in (C2), an idea which can be extended to any other search rate distribution (Dennis 1989). Model (C1) fits well to mating efficiency for many invertebrates (azuki bean weevil *Callosobruchus chinensis*; Dennis 1989, gypsy moth; Tcheslavskaia

Table 3.2. Possible functional forms $P(M,F)$ for the female mating rate (in continuous-time models) or the probability of a female to mate within a time step (in discrete-time models); M and F denote male and female population density, respectively.

Functional form of $P(M,F)$	Main assumptions	Selected references	Code
$1-e^{(-M/\theta)}$	Male density does not vary with time (referred to as unlimited polygyny below in this table); all females search for males at an equal rate; males and females are randomly distributed in space and move at random until entering a mate detection range within which mating is certain (referred to as RAND). Here and in the subsequent formulas, θ is a measure of the Allee effect strength (the higher θ the stronger the mate-finding Allee effect)	Philip (1957), Klomp et al. (1964), Gerritsen and Strickler (1977), Kuno (1978), Dennis (1989), Hopper and Roush (1993), McCarthy (1997)	C1
$M/(M+\theta)$	Unlimited polygyny; exponentially distributed search rate of females; RAND. Alternatively, males are depleted with time due to pairing (referred to below in this table as monogamy); all females search for males at an equal rate; 1:1 adult sex ratio ($M = F$); RAND	Dennis (1989), Veit and Lewis (1996), McCarthy (1997), Wells et al. (1998)	C2
$(M\,e^{[(M-F)/\theta]}-M)/(M\,e^{[(M-F)/\theta]}-F)$	Monogamy; all females search for males at an equal rate; unbalanced adult sex ratio ($M \neq F$); RAND. For $\theta\to0$ this function reduces to $\min(M,F)/F$, for $M/F\to1$ it equals $M/(M+\theta)$ i.e. (C2) is a special case of (C3) with 1:1 sex ratio	Wells et al. (1990), McCarthy (1997)	C3
$1-(1-a)^{cM}$	Parameter a is probability that a particular male succeeds in mating with a particular female; fraction c of males is assumed active. If a is reparameterised as $a = 1-(1/2)^{1/A}$, then $P(M,F) = 1-(1/2)^{cM/A}$ where A/c is a measure of the Allee effect strength (= number of males required for half the females to mate)	Kuno (1978), Stephan and Wissel (1994), Wells et al. (1998), Grevstad (1999b)	C4
$1-(1-M/T)^n$	Females search for randomly distributed territorial males (T territories) by randomly sampling n territories per unit time (e.g. one reproductive season or year)	Lamberson et al. (1992)	C5
$(M\,e^{[(M-F/h)/\theta]}-M)/(M\,e^{[(M-F/h)/\theta]}-F/h)$	Each receptive male has an ability to mate with up to h females per unit time (limited polygyny); unbalanced adult sex ratio ($M \neq F$); RAND. For $\theta\to0$ (no Allee effect) this function reduces to $\min(hM,F)/F$ which is one of the most popular and agreed descriptions of the pair formation rate outside the realm of Allee effects. For $M/F\to1$ it equals $M/(M+\theta)$ (C2). Monogamy restored for $h = 1$ (C3) and model (C1) recovered for $h\to+\infty$	This book	C6
$1/(1+e^{(a-M/b)})$	Sigmoidal curve, $a > 0$, $b > 0$	This book	C7
$M^d/(M^d + \theta)$	Sigmoidal curve, $\theta > 0$, $d > 1$; analogy with type III functional responses	This book	C8
$(M+a\theta)/(M+\theta)$, 0 at $M=0$	Hyperbolical curve approaching a for $M\to0$, $\theta > 0$, $0 \leq a < 1$	This book	C9
$(M+a)/(M+\theta+a)$, 0 at $M=0$	Hyperbolical curve approaching $a/(\theta+a)$ for $M\to0$, $\theta > 0$, $a \geq 0$	This book	C10
$1-(1-a)e^{(-M/b)}$, 0 at $M=0$	Hyperbolical curve approaching a for $M\to0$, $\theta > 0$, $0 \leq a < 1$	This book	C11

et al. 2002, sea urchin *Paracentrotus lividus*; Vogel *et al.* 1982) and vertebrates (box turtle *Terrapene carolina*; Mosimann 1958, Atlantic cod; Rowe *et al.* 2004). Interestingly, the mate-finding Allee effect can also be invoked artificially by flooding a wild population with sterile conspecifics and so disrupting fertilization (Box 3.2; see also Section 5.1.4).

Other, more complex models of the mate-finding Allee effect also exist. Specific forms of $P(M,F)$ were developed to cover cases in which a male can achieve only

---

Box 3.2. **The sterile insect release technique used to control invasive pests**

In order to disrupt female fertilization and hence reduce pest population growth rate (see also Section 5.1.4), large number of sterile insects may be released. The efficiency of this technique has been much studied through population modelling. Simple models show that wild (i.e. fertile) populations can be eradicated provided that a sufficient number of sterile insects are maintained in the population or if sterile insects are pumped into it at a sufficient rate (Lewis and van den Driessche 1993). Where pests have already started to spread spatially via a travelling wave, the spread can be reversed and the pest made extinct provided that we keep numbers of sterile insects sufficiently large or supply them at a sufficiently high rate (Lewis and van den Driessche 1993).

The technique works because of the emergence of mate-finding Allee effect that is sufficiently strong to cause a demographic Allee effect in the insect population (Barclay and Mackauer 1980a, Dennis 1989). If there are M fertile and S sterile males in the population, and each female finds a mate and mates just once in a reproductive season, the probability of a female reproducing that season is $M/(M+S)$, the form (C2) of Table 3.2. Simply stated, the technique will work if it can guarantee a large enough S for the overall population to have a negative per capita growth rate, i.e. decline. Predators (or parasitoids) which specialize on the pest species may increase the efficiency of the technique by lowering the minimum rate of sterile insect release required for pest eradication (Barclay and Mackauer 1980b). Efficiency can also be enhanced by the action of a generalist predator with a type II functional response: it may create a predation-driven Allee effect (below) which in turn may interact with the mate-finding Allee effect in such a way that the Allee threshold due to the resulting double Allee effect is disproportionately larger than any of the Allee thresholds due to the single Allee effects (Section 3.2.2).

Figure 3.2. The gypsy moth, *Lymantria dispar,* is subject to a mate-finding Allee effect, which generates cyclical invasion dynamics. USDA APHIS PPQ Archive, Bugwood.org

a limited number of matings during a period of time and where either only virgin females mate or all females have an ability to mate many times (Kuno 1978). Also, inspired by mating biology of pelagic copepods, Kiørboe (2006) derived formulas that quantify the female mating rate provided that females (i) need to mate just once (i.e. have an ability to store sperm) or (ii) need to mate each spawning season, and males (i) have an unlimited mating capacity or (ii) have a limit on how many matings per unit time they can undertake.

All of the above-mentioned mate-finding Allee effect models share an implicit assumption that males and females search for each other in a homogeneous space in a more or less random fashion. A completely different situation arises, however, once the population is patchily distributed and males and females search for each other only within patches. This occurs, e.g., among primates forming social groups (Dobson and Lyles 1989) or among parasites patchily distributed within their hosts, such as the sheep tick *Ixodes ricinus* in sheep (Rohlf 1969), schistosomiasis and other helminthic infections in humans (May 1977a), and the Karnal bunt pathogen in wheat (Garrett and Bowden 2002). In these cases, the probability that a female mates with a male depends on (i) the distribution of different sexes or mating types among patches or the probability that at least one male and one female will be present in a given patch, (ii) the probability that the two meet within the patch (this event is often assumed to occur with certainty), and (iii) the mating system. The resulting form $P(M,F)$ is here a function of the average number of individuals per patch, sex ratio in the population as a whole, and distribution of individuals across patches; see the above-referenced studies for specific forms.

Whereas the $P(M,F)$ forms based on the assumption of spatial homogeneity have a hyperbolical shape akin to a type II functional response (Fig. 3.3C), forms based on assuming patchily distributed populations are sigmoidal, akin to a type III functional response (Fig. 3.3E) (see below for more on functional responses). These sigmoidal forms can occasionally arise in a homogeneous space, too, for example if there is facilitation or stimulation of mating in females by the presence of males. This seems to occur in the milk conch *Strombus costatus* and possibly also other congenerics, where females which are laying eggs (i.e. have already been mated) are more likely to copulate and lay (more) eggs again than non-spawning females; non-spawning females are more likely to occur at lower male densities (Appeldoorn 1988; for some other examples see Chapter 2). Stoner and Ray-Culp (2000) even observed that reproduction in the queen conch *S. gigas* ceased when adult density fell below 50 conch ha^{-1}, and attributed this to a failure to encounter mates in low-density populations. (Gascoigne and Lipcius (2004c) suggested that delayed functional maturity plus fishing of young adults might be an alternative explanation for the observed threshold in reproduction.) No $P(M,F)$ form has so far been proposed to describe such situations—we propose two simple sigmoidal forms as models (C7) and (C8) of Table 3.2.

A feature that might make all hyperbolical (but also sigmoidal) forms of $P(M,F)$ too restrictive is their behaviour close to zero male density. One male, the minimum to ensure at least some female fertilization, can formally correspond to any low density (depending on how we scale space) and even that single male can in theory fertilize any fraction of females (or broadcasted eggs or plant ovules). A specific example where functions of this type may be appropriate is provided by plants that allow for vegetative reproduction or selfing (Taylor *et al.* 2004, Morgan *et al.* 2005). If a sharp initial increase in $P(M,F)$ corresponding to the step from no male to one male is followed by a slow increase in $P(M,F)$ up to 1 as M/F further increases, existing $P(M,F)$ forms are not fully adequate since their behaviour around zero male density is constrained by their continuity property. A more adequate option in such cases could be to let $P(M,F)$ approach a value between 0 and 1 as M approaches zero (while still keeping $P(0,F) = 0$). Three heuristic functions satisfying this behaviour are suggested in Table 3.2 as models (C9)–(C11).

In summary, there are a great variety of models of the mate-finding Allee effect which all have to meet a number of generic properties. They can be relatively simple functions of male density, but also complicated functions incorporating the density of both males and females. The female mating rate may increase either hyperbolically or sigmoidally with increasing male density, and may or need not approach zero as male density declines.

Models of predation-driven Allee effects
Allee effects were long considered to occur mainly or only in species with specific types of life history, especially those with particular modes of reproduction

(see Section 2.2). Later on, some factors external to populations were demonstrated to generate Allee effects. These are exploitation (Dennis 1989, Stephens and Sutherland 1999, Courchamp *et al.* 2006) and predation (Liermann and Hilborn 2001, Gascoigne and Lipcius 2004a), suggesting that Allee effects might be relevant to many more taxa than previously thought (see Section 2.3.2). Effects of an overall mortality rate $p(N)$ due to exploitation or predation can be modelled by extending equation (3.1) as

$$\frac{dN}{dt} = Ng(N) - p(N) \tag{3.4}$$

Exploitation and predation are in many respects the same, and exploitation has long been one of the major focuses for Allee effect models. Modellers usually distinguish constant effort exploitation whereby a population is exploited at a rate proportional to its size or density—$p(N) = EN$ for some $E > 0$—and constant yield exploitation whereby a population is exploited at a constant rate irrespectively of its size or density—$p(N) = E$ for some $E > 0$. Constant effort exploitation decreases individual fitness uniformly over the whole range of population sizes or densities. Therefore, it cannot create an Allee effect, but makes an existing Allee effect stronger (see Fig. 5.4). Constant yield exploitation, on the other hand, decreases individual fitness the more the smaller or sparser is the population, and thus generates a component Allee effect: the probability of an individual being captured in a given time increases as population declines (Dennis 1989, Stephens and Sutherland 1999). This apparently happened in the Norwegian spring-spawning herring fishery where the annual catch remained almost constant until the stock size was very small (see Fig. 5.7). For more on exploitation and Allee effects, and the relationship between exploitation and predation, see Section 5.2.

The relationship between Allee effects and full predator–prey dynamics is reviewed further on in this chapter (Section 3.6.1). Here we assume that the predator population does not respond numerically to changes in prey abundance. This situation best fits a generalist predator which is in a dynamic association with another (main) prey, and consumes the target prey as a secondary resource. This was the case in the island fox example, where golden eagles preying upon feral pigs also attacked foxes as the secondary prey (Section 2.3.2). Formally, this allows us to stay in the single-species framework.

To simplify things, let us first assume that there is no predator interference, i.e. consumption rates of individual predators sum up, and that there is no aggregative response of predators to prey, i.e. predator numbers in a given area do not respond to prey density. Functional response, the central concept to studies of predation, describes the number of prey each predator consumes per unit time as a function of prey population size or density. The most common form for this relationship, usually called a type II functional response, is a hyperbolical curve, which rises from zero to an asymptote (Fig. 3.3C). This form means that although there is an increase in predator consumption rate with prey density, this increase is not enough to offset the rate of increase in prey numbers. As a consequence, as prey density increases, there are more prey individuals per predator attack, and thus a lower probability that any prey individual will be eaten by a predator—a predation-driven Allee effect (Fig. 3.3D). Classical predator–prey theory further recognizes type I functional response (Fig. 3.3A) and type III or sigmoidal functional response (Fig. 3.3E). For a mathematical description of all three functional response types and likely examples of animals which obey them, see Table 3.3.

Interference among predators makes any functional response also a function of predator density, often through a functional response parameter. For example, searching efficiency of a predator may decrease and/or handling time of prey may increase with increasing predator density (Hassell 2000). On the other hand, predators may benefit from the presence of conspecifics—consider albatrosses searching for patchily distributed krill (Fig. 2.14) or African wild dogs hunting cooperatively for their ungulate prey. In such cases, searching efficiency might actually increase and/or handling time decrease with increasing predator density, at least at lower predator densities. In addition, predator density may vary due to an aggregative response, which describes the number of predators per unit area as a function of prey population size or density (Gascoigne and Lipcius

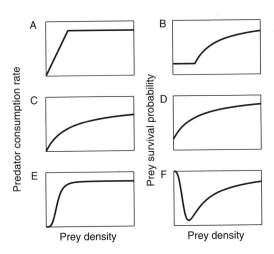

Figure 3.3. Predator consumption rates (left column) and corresponding prey survival probabilities (right column) for three classical types of predator functional response: type I functional response (A, B), type II functional response (C, D) and type III functional response (E, F). Redrawn from Gascoigne and Lipcius (2004a).

Table 3.3. Classical types of predator functional response. See Fig.3.3 for their graphical representation.

Functional response	Likely examples of predators	Equations used to model it	References
Type I	Aquatic suspension feeders, web or trap builders like spiders, humans	$f(N) = aN$ if $N < 1/(ab)$ and $f(N) = 1/b$ if $N \geq 1/(ab)$; $a>0$ scales the encounter rate with prey and $b>0$ the processing rate	Rigler (1961), Wilhelm *et al.* (2000), Crawley (1983), Martínez and Zuberogoitia (2001), Eggleston *et al.* (2003), McJunkin *et al.* (2005)
Type II (hyperbolical)	Considered most common. Predators of prey with passive predation avoidance strategies such as armouring or aposematic colouring, herbivores	$f(N) = aN/(1+ahN)$; $h>0$ is the time a predator requires to handle one prey individual	Holling (1959), Ivlev (1961), Gross *et al.* (1993)
Type III (sigmoidal)	Predators that switch prey, predators of cryptic or refuge-seeking prey or prey that are less active at low densities	$f(N) = a(N)N/(1+a(N)hN)$ with $a(N) = cN^{d-1}$, $c>0$, $d>1$, or $a(N) = cN/(1+dN)$, $c>0$, $d>0$	Murdoch and Oaten (1975), Hassell *et al.* (1976), Real (1977), Liermann and Hilborn (2001)

2004a). The consideration of interference, facilitation, and aggregative response thus results in a large number of functions describing predation-driven mortality of the target prey.

Whether we deal with a 'simple' functional response, a functional response combined with an aggregative response, or a functional response modified by interference or facilitation, the general conditions under which predation creates a component Allee effect in prey survival can easily be derived. Let $F(N) = 1 - [p(N)/N]\Delta t$ denote the probability that a prey individual escapes predation in a (small) time interval Δt. For a predation-driven component Allee effect to occur, we require $F(N)$ to increase with increasing prey population size or density (mathematically, $dF/dN > 0$ for all $N \geq 0$). This implies

$$\frac{dp(N)}{dN} < \frac{p(N)}{N} \text{ for all } N \geq 0 \qquad (3.5)$$

Substituting for $p(N)$ the functional response types listed in Table 3.3, this inequality holds for type II functional responses, but not for type I and type III functional responses. Figure 3.3 shows the form of $F(N)$ for each of these functional response types. Sigmoidal (type III) functional responses can result in a 'predator pit'—rare prey populations do not go extinct but rather approach a stable, low-density equilibrium maintained by predation which is potentially far below the prey carrying capacity (May 1977b); we return to this point in Chapter 6.

In some species it is prey behaviour rather than predator consumption of prey which is a mechanism for an Allee effect in prey—consider a herding, flocking or schooling species in which efficiency of anti-predator behaviour decreases with declining prey group size (Section 2.3.2). Even in this case, however, prey dynamics might be analogous to those predicted by models with a type II functional response. Consider, for example, a predator with a linear functional response $f(N) = aN$ with a positive attack rate a. Since the efficiency of anti-predator behaviour increases with increasing prey abundance N, the predator attack rate decreases and may be described, e.g., as $a = \alpha/(\beta+N)$. By combining these two expressions, we get $f(N) = \alpha N/(\beta+N)$ which is just another parameterization of a type II functional response (although the parameters involved have a different meaning; see Table 3.3). This implies that models of predation-driven component Allee effects might, at least in some cases, be used to describe component Allee effects due to reduced efficiency of anti-predator behaviour in small prey populations.

Stock-recruitment models in fisheries

Because of the difficulty of detecting and counting fish and other marine organisms smaller than a threshold size, the fisheries literature almost invariably subsumes reproduction of adults and survival of the first few life stages jointly into a stock-recruitment (SR) model. SR models relate the number of spawning individuals or stock size or density S to the number of recruits R surviving up to a given time after hatching, usually the time needed for the young to achieve the threshold size for quantitative sampling.

The two most commonly used SR models, the Beverton-Holt model $R = aS/(1+bS)$ and the Ricker model $R = aS\,e^{-bS}$, are negatively density dependent—the per spawner recruitment R/S decreases as the stock size or density S increases—and a trick is needed to give them a hump-shaped form if an Allee effect in reproduction and/or early survival is to be modelled. The most popular and flexible SR model that demonstrates an Allee effect (or depensation, as it is usually termed in fisheries[1], by analogy with compensation as a term for negative density dependence) appears to be the modified Beverton-Holt model (Myers *et al.* 1995, Liermann and Hilborn 1997, Gascoigne and Lipcius 2004b)

$$R = \frac{aS^d}{1+bS^d} \tag{3.6}$$

---

[1] Arguably depensation is a better term than 'Allee effects' but we have opted for the term in more widespread use in the ecological community.

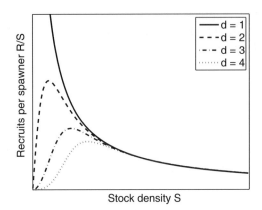

Figure 3.4. Per spawner recruitment as a function of stock density in the modified Beverton-Holt stock-recruitment model (3.6). Parameters: $a = 1000$, $b = 45$; note that $d = 1$ corresponds to the null case of no Allee effect. Redrawn from Gascoigne and Lipcius (2004b).

where a and b are positive constants and $d > 1$ controls the strength of the Allee effect (Fig. 3.4; $d = 1$ gives back the original Beverton-Holt model). Alternatively, one may introduce an Allee effect into any (negatively density-dependent) SR model through multiplying it by the term $S/(S + d)$ (Barrowman *et al.* 2003). Here again, $d > 0$ controls the strength of the Allee effect; the original SR model is recovered in this case for $d = 0$. Finally, SR models can be made positively density-dependent by replacing the stock size or density S with the difference $S - S_c$ and setting recruitment to zero once the stock size drops below a critical stock size or density S_c (Franck and Brickman 2000).

SR models are not mechanistic but can be adequate and sufficiently flexible to fit a number of stock-recruitment data sets. In fact, data fitting represents one of the most common applications of these models—they have been used to successfully detect Allee effects in such species as the coho salmon, *Oncorhynchus kisutch* (Chen *et al.* 2002) and the Pacific sardine, *Sardinops sagax* (Morales-Bojórquez and Nevárez-Martínez 2005) and reject the hypothesis of presence of an Allee effect in a number of other species (Myers *et al.* 1995, Chen *et al.* 2002).

Other models of component Allee effects
There are also other models of component Allee effects. Given the variety of Allee effect mechanisms, it is not surprising that component Allee effect models also vary markedly. They range from generic statistical models used to fit observed data to very specific models tailored to the particular component Allee effect under study. The three examples given in Box 3.3 demonstrate how various models of component Allee effects can look.

Box 3.3.   **Examples of specific approaches to modelling component Allee effects**

Alpine marmots (*Marmota marmota*) live in social groups of up to 20 individuals, and hibernate communally over the winter. By modelling the dynamics of the alpine marmot population in the Berchtesgaden National Park, southern Germany, Stephens *et al.* (2002a) aimed to determine whether the probability of an individual marmot surviving winter depends on group size. They fitted a statistical model

$$1 - \frac{1}{1 + e^{c_1 q_1 + c_2 q_2 + \dots}} \tag{3.7}$$

to empirical data; c_i are coefficients that scale factors q_i affecting winter survival, such as winter length, age, numbers of adults and yearlings in the group etc. Positive effects of adult and/or yearling numbers were demonstrated in three out of five marmot stage classes[2] considered. The component Allee effect in winter survival was hypothesized to be a consequence of less efficient social thermoregulation during hibernation in small marmot groups.

The fruit fly *Drosophila melanogaster* oviposits on decaying fruit. Increased adult densities on the fruit prior to larval development yielded higher larval survival, thought to be because adult flies inoculated the fruit with yeasts which in turn reduced fungal growth detrimental to developing larvae (Wertheim *et al.* 2002). Etienne *et al.* (2002), developing a spatio-temporal model of *D. melanogaster* population dynamics, formalized this observation by relating the number of larvae hatching on a piece of fruit (*L*) to the number of adult females emerging from that fruit (*A*) in a rather crude way:

$$A = L/2 \quad \text{if } L_{min} \le L \le L_{max} \text{ and } A = 0 \text{ otherwise} \tag{3.8}$$

Here the lower bound for adult emergence L_{min} is set by a component Allee effect due to environmental conditioning (minimum number of adults necessary to sufficiently mitigate the impact of fungi). The upper bound L_{max} stems from the observation that high numbers of larvae are exposed to severe scramble competition for the fruit content (Etienne *et al.* 2002, Wertheim *et al.* 2002). Larval fitness is thus maximized at an intermediate larval abundance. An alternative, statistical model describing a component

---

[2] Stage classes are similar to year classes but of less well defined length; an example would be pup vs. juvenile vs. adult. For more on age- and stage-structured models, see Section 3.2.3.

Box 3.3. (*Continued*)

Allee effect in larval survival of *D. subobscura* was proposed by Rohlfs and Hoffmeister (2003).

Estuaries on the US Pacific coast currently face an invasion of the wind-pollinated smooth cordgrass (Davis *et al.* 2004, Taylor *et al.* 2004). Individuals initially establish by seeds, then grow vegetatively to produce circular patches of clonal plants that can extend several metres across. As the invasion proceeds, the formerly isolated clones coalesce and form continuous meadows. Members of the *Spartina* population can thus be classified according to their local density, as seedlings, isolated clones, or meadow-forming individuals. *Spartina* is largely self-incompatible—clones are limited by pollen from their distant conspecifics and suffer from a reduced seed set relative to meadow-forming individuals (Davis *et al.* 2004). Therefore, classification of population members according to their local density into a finite number of categories allowed for an easy modelling of the component Allee effect due to pollen limitation—clones were simply assigned a lower fecundity (seed set) than meadow-formers.

Figure 3.5. The smooth cordgrass, *Spartina alterniflora*, spreads through isolated clones, which show a component Allee effect due to pollen limitation © Galen Rowell/Corbis.

3.2.2. **Insights from simple population models**

How can one determine whether a component Allee effect gives rise to a demographic Allee effect? Section 5.3.3 suggests a number of approaches, including one based on modelling population dynamics which we discuss here in more detail. In order to assess the consequences of a component Allee effect for population dynamics, it is perhaps most instructive to start with embedding models of component Allee effects in a simple population model. Consider a (continuous-time) model

$$\frac{dN}{dt} = Nb(N) - Nd(N) \tag{3.9}$$

where $b(N)$ and $d(N)$ are (positively or negatively) density-dependent per capita birth and death rates, respectively. Models of this kind can be adequate, e.g., if detailed knowledge about the population is not available or as a raw test of whether an identified component Allee effect is strong enough to generate a demographic Allee effect. Consider a specific version of model (3.9) which describes the dynamics of a population subject to two component Allee effects (Berec *et al.* 2007):

$$\frac{dN}{dt} = \underbrace{b\left[1 - (1-a)\exp(-N/\theta)\right]N}_{Reproduction} - \underbrace{d\left(1 + \frac{N}{K}\right)N}_{\substack{Mortality\ from \\ natural\ causes \\ (other\ than\ predation \\ by\ the\ focal\ predator)}} - \underbrace{\frac{\alpha N}{1+\beta N}}_{\substack{Predation-driven \\ mortality}} \tag{3.10}$$

Divided by N, all terms are per capita, and the three terms on the right-hand side represent in sequence (i) a positively density-dependent birth rate (mate-finding Allee effect; model C11 of Table 3.2), (ii) a negatively density-dependent survival rate due to factors other than predation, and (iii) a positively density-dependent survival rate owing to predation (predation-driven Allee effect; type II functional response of Table 3.3). Positive constants $a < 1$ and θ define the intensity of the mate-finding Allee effect, b is the maximum birth rate, d is the mortality rate at low densities and in absence of the predation-driven Allee effect, $K > 0$ scales the carrying capacity of the environment for prey, and positive constants α and β scale the predation rate. Because setting $a = 1$ and $\beta = 0$ switches off the mate-finding and predation-driven Allee effect, respectively, we can assess the implications of both component Allee effects for prey dynamics in isolation and also simultaneously.

Demographic Allee effects readily arise due to reduced mating efficiency in low-density populations (Fig. 3.6A). The Allee threshold increases with the death-to-birth-rate ratio d/b; for any fixed d/b, the Allee threshold increases as a, the notional female mating rate at zero population density, declines. A straightforward implication of this simple modelling exercise is that any decrease in a (e.g. through evolution—Chapter 4) and/or in d/b (e.g. through protection from exploitation or regular augmentation—Chapter 5) mitigates (or entirely prevents) the negative impact of the mate-finding Allee effect on population dynamics. Conversely, in an exploited population with higher d and thus d/b, the Allee threshold increases, or one may appear where previously none was present (Fig. 3.6A; see also Fig. 5.4 for the same phenomenon in the context of fisheries).

Predators with a type II functional response have the potential to create demographic Allee effects in prey (Fig. 3.6B). There is no demographic Allee effect

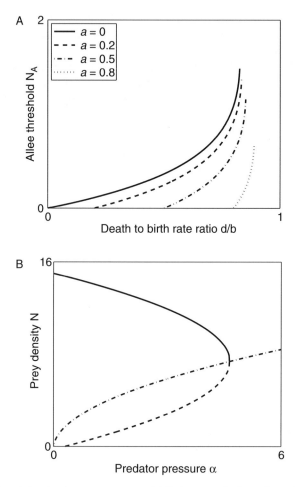

Figure 3.6. Population-level consequences of (A) the mate-finding Allee effect and (B) a predation-driven Allee effect, as predicted by model (3.10). A. Location of the Allee threshold (unstable equilibrium). For large enough d/b ratio the prey population goes extinct from any initial density. B. Locations of equilibrium densities of prey (black solid line = locally stable/carrying capacity, black dashed line = unstable/Allee threshold) and densities at which the per capita prey growth rate is maximized (dash–dot line). There is no Allee threshold at low predator pressure and the prey population goes extinct from any initial density at high predator pressure. Model parameters: A. $\alpha = 0$, $K = 10$, $\theta = 0.5$; B. $a = 1$, $K = 10$, $b = 0.5$, $d = 0.2$, $\beta = 4$.

when predation is weak, but as predator density (or more generally, the parameter α) increases, a demographic Allee effect appears, first weak, then strong. Once α is high enough, predation becomes fatal—the prey population becomes extinct whatever its initial density. Management of natural enemies may thus considerably mitigate the negative impact of Allee effects on prey population.

The ways in which the two component Allee effects interact are far from trivial. The outcomes of the model vary depending on the values of the parameters determining the strength of the individual component Allee effects (Fig. 3.7). To distinguish weak and strong interactions, model outcomes can be classified according to whether the overall Allee threshold is higher or lower than the sum of the two Allee thresholds corresponding to single component Allee effects; we call such cases superadditive and subadditive Allee effects, respectively, noting that superadditivity is in fact a synergistic interaction (Berec *et al.* 2007; see also Box 1.1). Of special interest are the cases in which the single Allee effects are weak, but the double Allee effect is strong; we then speak of dormant (or double dormant) Allee effects. A dormant Allee effect also occurs when one of the single Allee effects is weak, one is strong, and the Allee threshold due to the double Allee effect is higher than that of the strong single Allee effect. We use the word 'dormant' here to point out that although a weak Allee effect is rarely a reason for concern when alone, it may significantly increase the threat of population extinction when interacting with another demographic Allee effect. The occurrence of dormancy implies that even weak Allee effects represent a risk that should be accounted for: should another Allee effect occur, for example through anthropogenic activities (Section 5.2), the population could go extinct much faster than would be expected from the Allee effect alone. Interactions of two or more component Allee effects cannot be disregarded in population management (Berec *et al.* 2007; Table 2.4).

Interestingly, while model (3.10) with $a = 1$ is almost invariably interpreted in terms of predation and predation-driven Allee effects, Thieme (2003) derived it in the context of mate search and hence the mate-finding Allee effect, using a number of mechanistic arguments about searching for mates and mating. This implies that, as long as model parameters are given a proper interpretation, these two processes can be studied in a unified framework.

3.2.3. **State-structured population models**

Although simple, 'unstructured' models are sufficient to capture and understand dynamics of some populations, in others, population structure must be taken into account, since it often affects population growth rate and size in a significant way. How do structured population models look, and how can Allee effects be incorporated?

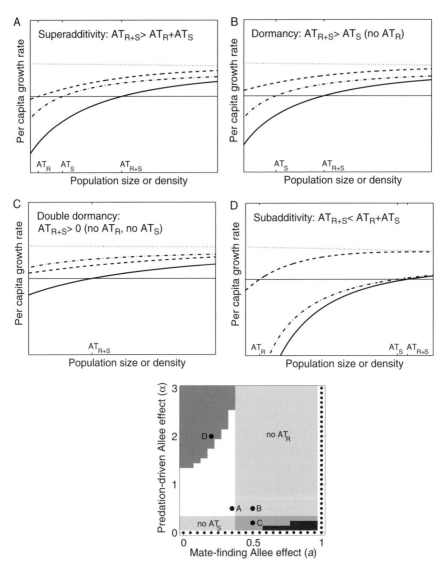

Figure 3.7. Population-level consequences of the interaction of two component Allee effects, as predicted by model (3.10). Panels A to D demonstrate how the per capita population growth rate depends on population size or density in the vicinity of Allee thresholds corresponding to single (AT_R and AT_S) and double (AT_{R+S}) Allee effects. Loosely speaking, the double Allee effect can be disproportionately stronger than any of the single Allee effects (superadditivity, A), can be stronger than one of the single Allee effects where the other is only weak (dormancy, B), can be strong even when both single Allee effects are weak (double dormancy, C) or can be only marginally stronger than a single Allee effect (subadditivity, D). Precise definitions of the terms are given symbolically in the respective panels A to D and in words in Box 1.1 of Chapter 1. Panel E shows the areas of models parameters characterizing strengths of the single Allee effects in which these patterns occur. Legend to panels A to D: dotted line, no Allee effect;

One way of classifying structured population models can be into (i) state-structured models, (ii) simulation models, (iii) spatial models and (iv) multiple-species models. This classification is also convenient from the Allee effects viewpoint. Whereas models of component Allee effects often act as building blocks for state-structured and simulation models, (phenomenological) models of demographic Allee effects are often a part of spatial and multiple-species models. As we are unable to discuss every structured population model that has ever been used to study Allee effects, our discussion of the single-species classes of models (i–iii) is supported by a table giving a more complete, but by no means comprehensive, list of theoretical studies related to Allee effects (Table 3.4). We consider state-structured population models in this section and simulation models in the next one; spatial models and multiple-species models are discussed in sections 3.5 and 3.6, respectively.

An initial impetus for developing structured population models was to understand a population's age distribution and its temporal dynamics. Age-dependent birth and death rates, widely recognized among populations and summarized in life tables, became an integral part of matrix population models (discrete time, discrete age classes) developed by P. H. Leslie in the 1940's, as well as of the McKendrick-von Foerster model that treats both time and age in a continuous fashion (Kot 2001). Later on, population models were structured in many other ways, including by sex, body size, developmental stage, density or any combination of these, and the matrix approach originated by Leslie has come to dominate structured population modelling (see Caswell 2001 for an introduction to these models). State-structured models (we refer to state to mean all these structural types) also exist which treat time as continuous and states as belonging to a finite set of classes (e.g. Ashih and Wilson 2001, Chapter 11 in Thieme 2003, Berec and Boukal 2004); mathematically, these are systems of ordinary differential equations.

In principle, any state-structured model can be used to study Allee effects. Various researchers studied Allee effects via population models structured with respect to sex (Hopper and Roush 1993, Ashih and Wilson 2001, Berec *et al.*

---

dashed line, mate-finding Allee effect; dash–dot line, predation-driven Allee effect; solid line, double Allee effect. AT_R, AT_S and AT_{R+S} represent the Allee threshold owing to the mate-finding Allee effect, predation-driven Allee effect and double Allee effect, respectively. In panel E, dots along the bottom and up the right-hand side represent the combinations where there is only either a single or no component Allee effect, white colour corresponds to superadditivity, light grey to dormancy, mid grey to double dormancy and dark grey to subadditivity; the black colour corresponds to the cases where two weak Allee effects combine to produce a joint weak Allee effect, and where a weak Allee effect and a strong Allee effect combine such that the overall Allee threshold equals that of the strong Allee effect. Model parameters of $b = 0.5$, $\theta = 0.5$, $d = 0.2$, $K = 10$ and $\beta = 4$ were chosen to represent a wide spectrum of possible interaction outcomes. Redrawn from Berec *et al.* (2007).

Table 3.4. An overview of a variety of Allee effect studies that used structured population models.

Model type	Objective	Main results	References
Strategic Structured by sex and space Mate-finding Allee effect	Do introductions of parasitoids fail due to a mate-finding Allee effect?	The critical number of females needed for establishing a founder population decreased hyperbolically with increasing mate detection distance and/or net reproductive rate, and increased linearly with mean-square displacement per generation. When the assumption that virgin females produce all males (arrhenotoky) was changed to virgin females producing no progeny, the critical number of females increased by over 30%.	Hopper and Roush (1993)
Strategic Structured by sex, space and female fertilization status Mate-finding Allee effect	Effects of gestation time on two-sex population dynamics.	Species with longer gestation times suffered a stronger Allee effect and were thus more vulnerable to extinction. More specifically, species with longer gestation times had a higher Allee threshold, a lower carrying capacity and a lower threshold mortality above which populations go extinct from any initial density.	Ashih and Wilson (2001)
Strategic Structured by body size Allee effect in juvenile survival and immigration	Effects of future harvesting strategies on population dynamics of the red sea urchin *Strongylocentrotus franciscanus*.	Even relatively low rates of fishing mortality resulted in a large (> 50%) decline in population size over 100 years if the Allee effect was included in the model. Although a rotational fishery had lower yields than annual exploitation, it caused red sea urchin populations to decline at a lower rate.	Pfister and Bradbury (1996)
Strategic Structured by developmental stage Allee effect in reproduction	Impact of delayed functional maturity, Allee effects in reproductive behaviour and stage-dependent fishery on vulnerability to population collapse in the queen conch *Strombus gigas*.	Either Allee effects in reproduction, or delayed functional maturity plus fishing of young adults resulted in a rapid collapse to extinction above some threshold mortality. Queen conch found vulnerable to population collapse under heavy fishing pressure. Difficulty of distinguishing different potential mechanisms for population collapse in the field.	Gascoigne and Lipcius (2004c)

Strategic Structured by sex, space and individual Mate-finding Allee effect (emergent)	How diverse mate search strategies affect population dynamics of sexually reproducing species	A hyperbolical extinction boundary in the male and female density space delimits populations going extinct (below the boundary) from those that persist (above the boundary). The area of extinction shrank as mate search strategy changed from random to local and from passive to active, and, in the active case, as the mate detection distance increased.	Berec *et al.* (2001)
Strategic Structured by sex, space, mating status and individual Mate-finding Allee effect (emergent)	How mate search strategies, degree of mate choice and degree of mate fidelity affect population dynamics of sexually reproducing species	Individuals with higher divorce rates cannot afford to be as choosy if the population is to persist. Individuals can afford both higher choosiness and lower mate fidelity as mate search strategy changes from passive to active. Longer-lived species persist at higher degrees of mate choice and lower degrees of mate fidelity as compared with shorter-lived ones.	Berec and Boukal (2004)
Tactic Structured by sex, space, developmental stage and individual Allee effect due to environmental conditioning and mate-finding Allee effect (emergent)	A data-based comparative study of four population models of increasing complexity—do they succeed in predicting population dynamics of the alpine marmot *Marmota marmota*?	Dynamics predicted by a behaviour-based model (simulation model with optimised habitat selection behaviour) and two matrix population models were comparable to those observed in the field, but dynamics of a fully probabilistic simulation model without optimised behaviour were not. The behaviour-based model was best at predicting transient dynamics, and also predicted the strongest demographic Allee effect of all models examined. Considering realistic patterns of behaviour in spatially explicit models may thus be vital for understanding the dynamics of a specific population.	Stephens *et al.* (2002a)
Strategic Structured by space and developmental stage Allee effect due to environmental conditioning	Population-dynamic implications of an Allee effect, scramble competition and dispersal in the fruit fly *Drosophila melanogaster*.	Initial distribution and density of adults determined whether a population could become established, but resource availability affected subsequent population persistence. Relative to straightforward diffusion, metapopulation persistence was facilitated by a dispersal mode characterized by higher probabilities of travelling both short and long distances and smaller probabilities of travelling intermediate distances.	Etienne *et al.* (2002)

Table 3.4. (Continued)

Model type	Objective	Main results	References
Tactic Structured by sex, space, age and social status Allee effect due to cooperative breeding	Spatial dynamics of populations of the red-cockaded woodpecker *Picoides borealis*, an endangered and territorial cooperative breeder.	Population stability most sensitive to mortality of female breeders and female dispersers, and to the number of fledglings per brood. As territories got aggregated, a lower number of territories sufficed for the population to be stable.	Letcher *et al.* (1998)
Tactic Structured by sex, space, age, social status and individual Allee effect due to cooperative breeding	Vulnerability of populations of the red-cockaded woodpecker *Picoides borealis* to demographic end environmental stochasticity.	Because presence of helpers can ameliorate the impact of stochasticity in mortality and reproduction on the size of the breeding population, vulnerability of woodpeckers to these threats is relatively low. Spatially restricted dispersal of helpers makes this effect most pronounced when territories are aggregated or at high densities. As a result, relatively small populations can still be relatively stable, as long as they are dense.	Walters *et al.* (2002)
Strategic Structured by space and pollination status Allee effect due to pollen limitation	How spatial scale of pollen and seed dispersal affect dynamics of plant populations and strength of the pollen-limitation Allee effect.	The strength of the demographic Allee effect decreases as the spatial scale of pollen and/or seed dispersal decreases (populations are dramatically more clumped when colonization is local and less wasteful of pollen when pollination is local), or as individuals produce pollen and/or seeds at increasing rates. Strong Allee effects can even be avoided altogether for some combinations of these two mechanisms.	Stewart-Cox *et al.* (2005)
Tactic/strategic Structured by space and age Allee effect due to decreased fertilisation efficiency and reduced recruitment at low (adult) densities	Are marine reserves an effective conservation mechanism for marine metapopulations with an Allee effect? The model is applied to the red sea urchin *Strongylocentrotus franciscanus*.	Under some conditions, reserves could be instituted with little or no penalty in sustainable catch at low to moderate levels of fishing effort. The establishment of a reserve was necessary to prevent collapses and maintain sustainable catches at high levels of fishing effort. Multiple reserves, spaced more closely than the average larval dispersal distance, were most successful at maintaining healthy populations and sustainable levels of harvest.	Quinn *et al.* (1993)

Strategic Structured by developmental stage (juveniles and adults) Allee effect affecting reproduction	Population dynamics of a two-stage population with a component Allee effect.	If the Allee effect is absent populations attain their carrying capacity. Otherwise, periodicity arises out of a tension between the critical densities of each stage when the initial density of adults is above their respective critical value and the initial density of juveniles is below their critical value. Periodic dynamics are more frequently observed as adult mortality increases, up to a point beyond which they are replaced by extinction. This emergent periodicity cannot arise in an unstructured version of the model.	Gascoigne and Lipcius (2005)
Strategic Structured by space Generic strong Allee effect in local populations	What is the optimum number and size of reserves to protect metapopulations with a strong Allee effect at the local population level?	Given a negative relationship between the number of reserves and their mean size, there is an intermediate reserve number and size at which the mean time to metapopulation extinction is maximized provided that the Allee effect is not overly strong. Metapopulations go always extinct once a critical Allee effect strength is exceeded. The precise value of the optimum depends on the total habitat size, local population extinction rate, dispersal rate and the Allee effect strength.	Zhou and Wang (2006)
Strategic Structured by space Generic strong Allee effect in local populations	How are oscillatory dynamics in isolated populations with an Allee effect affected by between-population dispersal?	An increase in between-patch dispersal causes oscillations in each patch to decrease in amplitude. However, this 'stabilization effect' may be accompanied by an increase in extinction vulnerability of the metapopulation as a whole. Finite metapopulations in which some individuals leaving the boundary patches are lost to the outer environment go extinct from any initial state if the dispersal rate exceeds a critical value.	Hadjiavgousti and Ichtiaroglou (2006)
Tactic Structured by age and sex A compound component Allee effect in fecundity due to several sources	Is there a threshold density below which populations are likely to collapse due to a demographic Allee effect? Model is applied to the samango monkey *Cercopithecus mitis*.	Populations of samango monkeys cannot tolerate more than a 60–70% decrease across all age-sex cohorts from the equilibrium-density troop structure. Populations experiencing less than a 60% decrease in density due to some catastrophic event recover rapidly to equilibrium levels provided the carrying capacity of the environment is unaffected.	Swart *et al.* (1993)

2001, Berec and Boukal 2004), age (Cushing 1994, Kulenovic and Yakubu 2004), body size (Pfister and Bradbury 1996, de Roos *et al.* 2003), developmental stage (Gascoigne and Lipcius 2004b, 2004c, 2005), and density (Taylor and Hastings 2004, Taylor *et al.* 2004)—see Table 3.4 for more details on some of these studies. Discrete time, state-structured models share a common matrix form

$$N_{t+1} = A(N_t)\, N_t \tag{3.11}$$

by which the current state distribution N_t (such as numbers of individuals in each age class) is updated to the state distribution in the next time step N_{t+1} by means of a (density-dependent) transition matrix A. Elements of this matrix are generally (positively or negatively) density-dependent functions which determine probabilities or amounts of inter-state transitions (such as the probability of an individual surviving to the next age class or the fecundity of an individual of a given age). Continuous-time, state-structured models have a form

$$dN\,/\,dt = A(N)\, N \tag{3.12}$$

in which the matrix A now determines (positively or negatively density-dependent) state transition rates. Component Allee effect models enter both these (matrix) equations as positively density-dependent or hump-shaped (ingredients of) matrix elements, such as stock-recruitment model with an Allee effect in Gascoigne and Lipcius (2005) or positively density-dependent probability of surviving from one stage to the next in Pfister and Bradbury (1996).

In state-structured population models one is no longer working with population size or density as a single number, but rather with a whole distribution of sizes or densities across population states (such as densities of males and females in a sex-structured population). This inevitably changes the perception of the Allee threshold as understood in all one-equation models discussed above. In higher-dimensional state-structured models (such as a two-dimensional sex-structured model) it is the combination of subpopulation sizes or densities that decides on population survival or extinction, not the size or density of the population as a whole. Schreiber (2004) showed that for a broad class of state-structured population models with a component Allee effect one can observe at most three possible dynamics: (i) population extinction from any initial size or density, (ii) population persistence for any initial size or density, and (iii) bistable dynamics akin to a strong Allee effect. For (iii), he further proved that there exists a boundary in the space of subpopulation sizes or densities which divides this space into two exclusive parts; for points below this boundary the population goes extinct and for points above it the population persists (Fig. 3.8). This very important result says that such an 'extinction boundary' is a generalization of the concept of the Allee threshold in state-structured population models.

Schreiber's result does not say anything about shape of the extinction boundary other than that it is smooth and bounded (in the mathematical sense). Its specific shape has to be revealed by simulations in each particular instance, as exemplified by Fig. 3.8C based on the two-dimensional, stage-structured model developed by Gascoigne and Lipcius (2005). The class of models for which an extinction boundary exists is broader than that considered by Schreiber, however. Extinction boundaries in sex-structured population models were studied by Berec *et al.* (2001) and

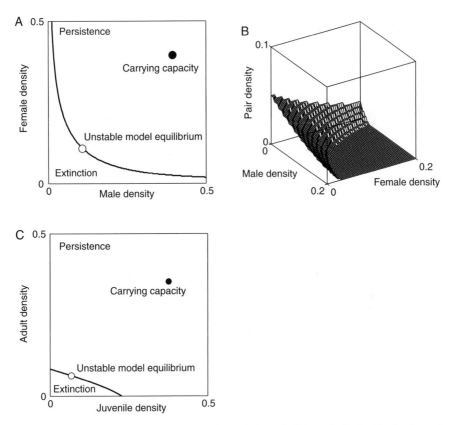

Figure 3.8. Allee effects in state-structured populations. A. A hyperbolical extinction boundary delimits areas of population extinction and persistence in male and female density space in a two-sex population model with the mate-finding Allee effect (Berec *et al.* 2001). B. A (two-dimensional) extinction boundary delimits areas of population extinction (below the surface) and population persistence (above the surface) in a three-dimensional, sex-structured model with the mate-finding Allee effect and pair bonds in which unpaired males, unpaired females and male-female pairs form mutually exclusive system states; redrawn from Berec and Boukal (2004). C. A bounded extinction boundary delimits areas of population extinction and persistence in juvenile and adult density space in a two-stage population model with an Allee effect in reproduction (based on the model developed by Gascoigne and Lipcius 2005).

Berec and Boukal (2004). Where only males and females were considered as the population states, the model predicted a hyperbolical extinction boundary splitting the space of male and female densities into regions of extinction and persistence (Fig. 3.8A). Where unpaired males, unpaired females and established male-female pairs formed three exclusive states (a necessary generalization for species such as geese, swans or albatrosses in which pair duration spans two or more reproductive seasons or even lasts for life), a two-dimensional extinction boundary in the three-dimensional density space now delimits regions of population extinction and persistence (Fig. 3.8B). Neither of these two extinction boundaries is bounded.

3.2.4. **Simulation models**

Simulation models are mostly population models which treat individuals as discrete entities and where population dynamics are no longer defined in terms of equations, but emerge from a set of behavioural and demographic rules repeatedly applied to each individual (Huston *et al.* 1988, Berec 2002, Grimm and Railsback 2005). The use of simulation models in studies of Allee effects is increasing (e.g. Berec *et al.* 2001, South and Kenward 2001, Etienne *et al.* 2002, Stephens *et al.* 2002a, Wiegand *et al.* 2002, Berec and Boukal 2004, Taylor *et al.* 2004—see also Table 3.4). Simulation models are often stochastic with many probabilistic rules (e.g. an event A occurs with a probability p and an event B with the complementary probability $1-p$). Due to their high versatility, simulation models can combine individual variation coming from a variety of sources, such as space and sex (e.g. South and Kenward 2001), space and density (e.g. Taylor *et al.* 2004), or sex, developmental stage and space (e.g. Stephens *et al.* 2002a). Although non-spatial simulation models also exist, most simulation models account for space and hence for local variation in density and local interactions among individuals.

The range of topics and modelling alternatives adopted by simulation models is vast. To give the reader an idea of how simple simulation models might look, we single out two examples here (Box 3.4) and leave some others for Table 3.4. Among other things, these two examples differ in the way component Allee effects are modelled (implicit vs. explicit), and in the character of questions they address (strategic vs. tactic).

Both models reviewed in Box 3.4 share a feature common to many other simulation models—model rules are probabilistic, that is, individual behaviour does not take variation brought about by local density into account. There are dangers of simulation models that do not incorporate realistic patterns of behaviour, however. Stephens *et al.* (2002a) showed that predicted marmot population growth was typically higher in a model where dispersal was context-dependent (rather than probabilistic based simply on age) because subordinates were less likely

---

Box 3.4. **Two examples of simple simulation models with Allee effects**

Firstly, we consider a study by South and Kenward (2001) who examined the effect of individual dispersal distance on the growth rate of a sexually reproducing population. The model developed in this study represented space as a lattice of hexagonal sites, with each site either vacant or occupied by an unpaired male, unpaired female, or a pair. Within each time step (one year), behavioural and demographic rules run as follows: Working through all unpaired adults in a random order, if there was a mate within a detection distance of an unpaired individual, the latter moved to the site occupied by the former, the two formed a pair, bred and produced a given number of off-spring. No mate within the detection distance meant no breeding that year. Each offspring was either male or female with equal probability. Yearlings dispersed independently away from their natal site in a random direction and for a negative-exponential-distributed distance. If the selected site was occupied by a pair or a same-sex individual, the disperser moved to the near-est unoccupied site within a relocation distance. If there was no such site, the disperser died. Following dispersal, annual mortality probability was applied to each individual. It took a number of time steps for a juvenile to become a reproductively mature adult.

The model contained no explicit formula relating an individual fitness component to population size or density in a positive way, yet it showed a demographic Allee effect as an outcome. Demographic Allee effects thus created are usually referred to as emergent Allee effects, since the under-lying component Allee effect is only implicit, here in the form of behav-ioural model rules—if local density is low enough, some adults simply fail to locate mates. At high dispersal distances, populations became diluted, attained low population densities and hence individuals had low mate-finding abilities. Therefore, the population growth was constrained by the inability of adults to find mates—the mate-finding Allee effect. In contrast, at low dispersal distances, local crowding and hence high mate-finding abil-ities, growth was constrained at high density by the inability of dispersers to find vacant territories. Population growth was highest in between these extremes. A similar modelling framework led Arrontes (2005) to conclude that the optimum dispersal strategy for the brown alga *Fucus serratus*, when invading coastal regions of northern Spain, is to disperse locally, with only a small fraction of propagules dispersing to some larger distances.

The second example considers modelling spatial spread of an invasive, wind-pollinated plant species; the smooth cordgrass (Taylor *et al.* 2004; see

---

Box 3.4. (*Continued*)

---

also Box 3.3). Space was arranged as a lattice of 1000x1000 square sites, each representing 1m² and either occupied by *Spartina* or vacant. Any member of the *Spartina* population was classified according to its local density as an isolated clone or a meadow-forming individual (individual adjacent to another individual). Each time step (one year), any vacant site adjacent to an occupied site had a probability of becoming occupied through vegetative growth. Each occupied site then produced a given number of seeds. As *Spartina* is largely self-incompatible, the mean seed set across clones (mainly self-pollination) was set lower than that across meadow-forming individuals (mainly outcrossing)—an explicit component Allee effect due to pollen limitation. Each seed was allowed to disperse in a random direction to a distance specified by an exponential distribution. Each seed that fell inside the modelled arena—the 1km² square—had a probability of becoming established, and established individuals turned into a clone or a meadow-forming individual the following year. No adult mortality was imposed on established individuals.

The demographic Allee effect (weak in this case as lone individuals were able to reproduce vegetatively or through self-pollination) was shown to slow down the spread of the plant in a Pacific Coast estuary relative to a hypothetical population with no Allee effect.

---

to disperse from smaller groups (where their presence had a strong effect on the survival of related juveniles) and more likely to disperse from larger groups (where they had a lower effect on expected survival). In spite of this, modellers often soldier on using entirely probabilistic population models in which individuals will often do silly things, just because probabilities dictate that they must. Having said that, where knowledge of individual behaviour is limited, simulation models with probabilistic rules can be usefully regarded as an exploratory null scenario.

3.3. Fitting Allee effect models to empirical data

Any of the above models can in principle be used to help us search for component and demographic Allee effects in empirical data. (Not all models are, however, equally useful for fitting.) Given long-term census data on population size or density, the most straightforward way to test for presence of a demographic

Allee effect in these data is to calculate and plot the (possibly log-transformed) per capita population growth rate (y-axis) versus (possibly log-transformed) population size or density (x-axis), and fit the resulting data pairs to a quadratic (i.e. hump-shaped) function (e.g. Pedersen *et al.* 2001, Dulvy *et al.* 2004, Angulo *et al.* 2007):

$$y = a + bx + cx^2 \tag{3.13}$$

This is actually a reparameterized version of model (P1) of Table 3.1. A demographic Allee effect is detected whenever $b > 0$ and $c < 0$, or, provided that the quadratic term is not statistically significant, if $b > 0$. There are, however, a few technical difficulties in thoughtlessly applying the simple linear regression to estimate the parameters a, b and c—see Box 5.2 for more details. On the other hand, given that there is an inherent bias towards detecting negative density dependence from census data (Freckleton *et al.* 2006) and that this bias is further exaggerated by measurement errors (Freckleton *et al.* 2006) and model oversimplification (Festa-Bianchet *et al.* 2003), any statistically significant detection of positive density dependence, even through simple linear regression, appears to be a strong evidence for demographic Allee effects.

Model (3.13) can likewise be used to test for the presence of component Allee effects (Angulo *et al.* 2007). A heuristic, factor-based function used by Stephens *et al.* (2002a) to search for a positive relationship between the probability of individual marmots surviving winter and marmot group size represents an alternative statistical model to use in search of component Allee effects (Box 3.3). Finally, as we emphasize in Section 3.2.1, model (C1) of Table 3.2 has often been used to fit data on mating efficiency as a function of population density. We also recall that data fitting represents one of the most common applications of stock-recruitment models in fisheries (Section 3.2.1).

3.4. **Allee effects in the world of stochasticity**

Except for the simulation models of Box 3.4, all the population models discussed so far have been deterministic. Although deterministic population models are generally much simpler to handle and analyse than their stochastic counterparts, they ignore one of the ubiquitous features of nature—its unpredictability. Random events affect dynamics of all populations, not excluding those subject to Allee effects. Population ecologists usually deal with two types of stochasticity, demographic and environmental (Engen and Saether 1998, Engen *et al.* 1998). Demographic stochasticity refers to chance variation in the number of individual births and deaths, and is generally thought to have an important effect on dynamics only in small or sparse populations, i.e. exactly the populations in which Allee effects are expected to operate. Environmental stochasticity, on the other hand, is a combined effect of forces external to the population and includes both irregular fluctuations of the abiotic environment (ranging from annual variation in climatic variables such as rainfall to catastrophic events such as floods, extreme droughts, fires or cyclones) as well as the other species (including anthropogenic pressure, predation, or disease epidemics), and affects the dynamics of populations of any size or density. How does stochasticity modify the operation of Allee effects?

The simplest way to incorporate environmental stochasticity into models is to let one or more model parameters (e.g. death rate) randomly fluctuate in time (e.g. good and bad years)—see, e.g., Steele and Henderson (1984) or Arrontes (2005). Stochastic population models that enable some sort of analysis can be classified as discrete-time or continuous-time branching processes (also referred to as Markov chains) or stochastic difference or differential equations (Engen and Saether 1998, Dennis 2002). Branching processes are akin to many simulation models in that they treat all individuals separately as discrete entities and make them behave according to a set of probabilistic rules (Dennis 1989, Lamberson *et al.* 1992, Stephan and Wissel 1994, Drake 2004, Allen *et al.* 2005). Stochastic equations often take the form of a deterministic population model to which a stochastic noise is added (Dennis 2002, Engen *et al.* 2003, Liebhold and Bascompte 2003).

To get an idea of how stochasticity (here demographic stochasticity) and demographic Allee effects combine to affect the dynamic properties of a population, we will study the stochastic counterpart of model (P8) with $g(N)$ of the form (P4) (Table 3.1). Derivation of the stochastic model (here a discrete-time branching process) follows Allen *et al.* (2005) and we refer to this study for technical details.

When it comes to stochastic population models, the three statistics most commonly used to evaluate population viability are (i) the extinction probability (i.e. the probability that the population goes extinct within a specified time interval), (ii) the mean time to extinction, and (iii) the first passage probability (i.e. the probability of attaining a large population size before attaining a small one; see below) (Drake and Lodge 2006). As regards Allee effects, we are interested in how these characteristics respond to initial population size.

In the deterministic world, the extinction probability has a switch-like character—it equals 1 below the Allee threshold (extinction) and is 0 above it (persistence). Under stochasticity, the extinction probability of a population subject to a strong Allee effect becomes a sigmoidally decreasing function of initial population size (Fig. 3.9E). The inflection point of this function corresponds to the Allee threshold (i.e. the unstable equilibrium) of the underlying deterministic model (Fig. 3.9E, Dennis 1989). As a consequence, in the stochastic world, any population which drops below the Allee threshold may still grow and persist (Fig. 3.9C), while any population which starts above it may nonetheless go extinct

(Fig. 3.9A). The further the population is from the Allee threshold, however, the lower is the probability of these two events occurring (Fig. 3.9E). This implies that under stochastic conditions, the Allee threshold can be defined as the point at which a population has an equal probability of extinction and persistence; a fact which is often used to locate the Allee threshold from empirical data on multiple populations, see Box 5.3. Populations which escape extinction tend to fluctuate around the carrying capacity of the environment given by the underlying deterministic model (Fig. 3.9A–D). The extinction probability of populations subject to weak Allee effects decreases exponentially with initial population size, but at a lower rate than for populations with no Allee effect (Fig. 3.9E).

Similar relationships hold for the first passage probability, which in the case of Fig. 3.9 has been defined as the probability that the population goes extinct before attaining the size of ten, a value arbitrarily designated to be well above the Allee threshold in this model (Fig. 3.9F). The mean time to extinction increases in a sigmoidal manner with the initial population size for strong Allee effects, and is higher and increases hyperbolically for weak Allee effects (Fig. 3.9G). The frequency distribution of the time to extinction is skewed to the left (Fig. 3.9H), an observation which appears to hold generally and which suggests that the mean is perhaps not the best summary statistic for extinction time (Allen *et al.* 2005).

These simulation results are rather robust to the way deterministic models are transformed into stochastic ones, the type of stochasticity (demographic, environmental, or both), the model type (branching process or stochastic equation, unstructured or state-structured model), and life history details (continuous or pulsed reproduction, overlapping or non-overlapping generations, polygamous or monogamous mating system) (Dennis 1989, 2002, Lamberson *et al.* 1992, Stephan and Wissel 1994, Engen *et al.* 2003, Liebhold and Bascompte 2003, Drake 2004, Allen *et al.* 2005, Drake and Lodge 2006). Many simulation models are in fact complex branching processes whose predictions only corroborate conclusions drawn from their simpler cousins (e.g. Berec *et al.* 2001).

3.5. **Allee effects in spatially structured populations**

Real populations are spatially extended and spatial population models have already become a common tool in population ecology. Recall that both simulation models described in Box 3.4 are also spatial; these models have been developed to study the effects of spatial variation within (local) populations occupying a relatively homogeneous patch of habitat. Scaling one level up, there are models which describe the dynamics of metapopulations; that is, collections of spatially separated, local populations connected by dispersal. Space is usually

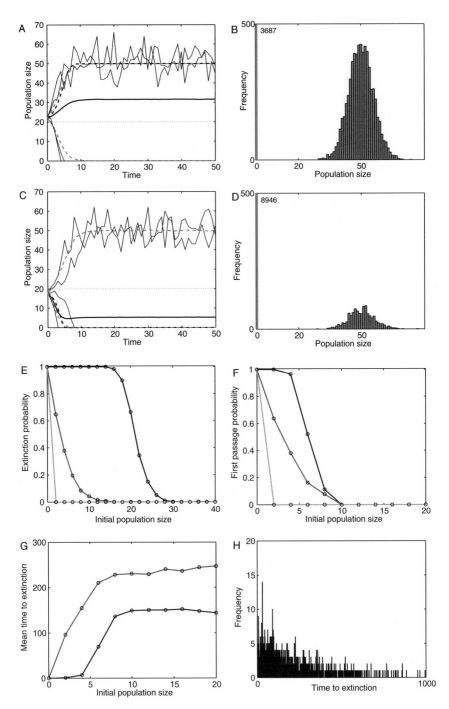

Figure 3.9. Impacts of (demographic) stochasticity on dynamics of a population subject to an Allee effect. Strong Allee effects cause the extinction probability and the probability of not attaining a large population size before attaining a small one to decrease sigmoidally with increasing initial population size (E and F, black), in contrast to weak Allee effects (E and

heterogeneous in this case, composed of a mosaic of inhabitable patches located within a sea of uninhabitable matrix. Metapopulation models are currently of high interest in conservation, since human-induced fragmentation of previously continuous natural habitats is ubiquitous (Section 5.1.1). However, metapopulation models can also be used to describe species in which there is a natural tendency to maintain spatially separated groups in an otherwise homogeneous environment, such as obligately cooperative breeders living in spatially separated breeding groups or fish living in spatially separated schools.

Spatial population models can also be viewed as strategic (conceptual) or tactic (realistic). More specifically, one may distinguish spatial models that have some kind of conceptual grid and allow (or not) individuals to leap from one cell to another in a single time step, from those models which have some kind of underlying GIS data layer and so consider habitat patches in terms of their real physical attributes and absolute distances. In this section, we discuss conceptual models of metapopulation dynamics, since these have almost exclusively been used to examine how Allee effects interact with space. In particular, we discuss three topics which have attracted the most attention from the Allee effect-oriented spatial modellers, and which are also of high importance for applied ecology: (i) source–sink dynamics, (ii) Allee-like effects in metapopulations and (iii) invasion dynamics.

F, grey) and no Allee effects (E and F, light grey) where the decrease is exponential. This implies that stochasticity may cause some populations starting above the Allee threshold to go extinct (A and B), while saving from extinction some populations that start below the Allee threshold (C and D). Mean time to extinction increases sigmoidally with increasing initial population size for strong Allee effects and results in a hyperbolical curve if the Allee effect is only weak (G). Finally, frequency distribution of the time to extinction is skewed to the left, which perhaps does not make its mean the best summarizing statistics (H). See the main text for more details.

Simulations were carried out for a stochastic version of model (P8) with $g(N)$ of the form (P4) as defined in Table 3.1. Model parameters and legend to the panels is as follows. A-D: $C = 1$, $A = 20$, $K = 50$, $r = 1.25$, 10000 simulation replicates. A+C: thin black—single trajectories, thick black—average over all simulations, dashed black—deterministic model prediction (A+B: initial population size = 22 > A, C+D: initial population size = 18 < A), dashed grey—averages over simulations leading to survival and extinction. B+D: frequency distribution of population sizes for all simulation replicates at time 50. E: extinction probability before time 50 for a strong Allee effect (black; $C = 1$, $A = 20$), weak Allee effect (dark grey; $C = 100$, $A = -2$), and no Allee effect (light grey; $C = 1$, $A = -1$), $K = 50$, $r = 1.25$, 10000 simulation replicates. F: probability of not attaining population size 10 before extinction as a function of initial population size, for a strong Allee effect (black; $C = 1$, $A = 5$), weak Allee effect (dark grey; $C = 100$, $A = -2$), and no Allee effect (light grey; $C = 1$, $A = -1$), $K = 15$, $r = 1.25$, 1000 simulation replicates. G: mean time to extinction for a strong Allee effect (black; $C = 1$, $A = 5$) and weak Allee effect (dark grey; $C = 100$, $A = -2$), $K = 15$, $r = 1.25$, 1000 simulation replicates. H: frequency distribution of extinction times for the initial population size = 10, $C = 1$, $A = 5$, $K = 15$, $r = 1.25$, 1000 simulation replicates.

3.5.1. **Source–sink dynamics**

Dispersal strategies are extremely important in determining the persistence of metapopulations, particularly when local populations are subject to strong Allee effects. The simplest spatial extension of non-spatial models is two-patch metapopulations. With Allee effects and without dispersal, both local populations persist only if population density in each patch is above a respective Allee threshold. Dispersal modifies this picture—two additional stable equilibria may appear in which one population persists below the Allee threshold and the other equilibrates above it (Fig. 3.10). This is possible because the low-density (sink) population is kept away from extinction by a steady inflow of individuals from the high-density (source) population (Gruntfest *et al.* 1997, Amarasekare 1998b, Gyllenberg *et al.* 1999). Below, we refer to these new equilibria as asymmetric, in contrast to the symmetric equilibrium in which both local populations are at their respective carrying capacity.

This basic picture has been further elaborated in various ways. Firstly, positive density dependence in dispersal rates (i.e. a lower propensity to leave a lower-density patch) facilitates persistence of populations below the Allee threshold (i.e. occurrence of source–sink dynamics) (Amarasekare 1998b). Secondly, both asymmetric and symmetric equilibria cease to exist once dispersal mortality gets too high (Amarasekare 1998b). This is an important point as regards human impacts—one result of increasing fragmentation is an increased distance between fragments and high dispersal mortality in matrix habitat which is frequently hostile (e.g. a matrix of agricultural land separating woodland patches). Thirdly,

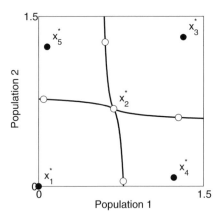

Figure 3.10.   Space of population densities in a two-patch system with source–sink dynamics in which there are four locally stable equilibria (black circles), two symmetric (x_1^* and x_3^*) and two asymmetric (x_4^* and x_5^*), and five unstable equilibria (white circles). The point x_2^* locates Allee thresholds corresponding to the two local populations 1 and 2. Solid lines delimit the areas of attraction of the respective stable equilibria. Modified from Gruntfest *et al.* (1997).

increasing dispersal rate weakens conditions for the occurrence of source–sink dynamics (Gruntfest *et al.* 1997). This may seem counter-intuitive since no sink can persist without dispersal, but effectively, increasing dispersal rate links the dynamics of local populations more and more, until they are basically one population. Fourthly, once there is sufficiently intense between-patch competition (as may happen e.g. in upstream vs. downstream populations of a suspension feeder) the asymmetric equilibria are the only stable equilibria, i.e. the symmetric equilibrium becomes unstable (Gyllenberg *et al.* 1999). Finally, active patch selection by dispersing individuals eliminates sink populations. That is, only the symmetric equilibrium and an equilibrium in which one population is at its carrying capacity and the other extinct are possible, as dispersers from other patches choose low-density patches with lower probability and hence put these patches at a greater risk of extinction (Greene 2003).

Although these results concern two-patch metapopulations, underlying mechanisms are general and should thus operate in multiple-patch systems, too. Only when all local populations fall below the Allee threshold (or, more precisely, below the extinction boundary as delineated in, e.g., Fig. 3.10) will the whole metapopulation start to decline. As an example, Frank and Brickman (2000) showed in the context of fisheries that even though all local populations (substocks) exhibited a strong Allee effect, it was possible for the metapopulation (stock) as a whole to appear negatively density dependent, with the consequences of (i) preventing detection of the Allee threshold at the local population level, and (ii) giving a false impression that fish populations could recover quickly from overexploitation if fishing pressure were relaxed. Under some circumstances, however, the metapopulation as a whole may persist even if all local populations are initially below their respective Allee threshold, provided that dispersal is asymmetric, i.e. some patches receive more immigrants than they lose (Padrón and Trevisan 2000). This raises the density of some local populations above their Allee threshold and these may persist permanently; some sinks are maintained through source–sink dynamics, and others go extinct.

Source–sink dynamics may also interact with stochasticity to prevent extinction of local populations. An isolated population close to the Allee threshold is likely to dip below it due to stochasticity and thus probably go extinct eventually. However, a local population in a metapopulation which dips below the Allee threshold can still be rescued. That is, the metapopulation structure provides a 'rescue' from stochasticity. Stochasticity also means that the identity of the source population and the sink population can vary over time.

3.5.2. **Allee-like effects in metapopulations**

The original metapopulation models were patch-occupancy models, formalizing the basic 'dogma' of metapopulation ecology: long-term metapopulation persistence occurs via a balance between two fundamental processes, extinction of local populations and colonization of empty patches of suitable habitat (Hanski and Gilpin 1997, Hanski and Gaggiotti 2004). Patch-occupancy models ignore local population dynamics, simply assuming that, in absence of an Allee effect, once a founder enters an empty patch then it takes a negligible amount of time for the local population to establish (i.e. to colonize the patch).

Amarasekare (1998a) extended the Levins patch-occupancy model of metapopulation dynamics by building upon its analogy with the logistic model of single-species population growth:

$$\frac{dp}{dt} = \left(cp(1-p) - ep\right)\frac{p-a}{1-e/c} \tag{3.14}$$

Here, p is the fraction of suitable patches that are occupied and c and e are the (maximum) patch colonization rate and the patch extinction rate, respectively. In addition, $0 < a < 1 - e/c$ is the threshold fraction of occupied patches below which the metapopulation as a whole goes extinct and above which it approaches the equilibrium fraction $1 - e/c$ of occupied patches. The threshold behaviour demonstrated by this model was termed the Allee-like effect, by analogy with Allee effects operating at the level of (local) populations (Amarasekare 1998a).

Mechanistic support for the idea of Allee-like effects comes from a modelling study of a metapopulation of the highly endangered African wild dog, a species that lives in spatially separated packs and needs a minimum pack size to successfully reproduce—a socially mediated strong Allee effect at the pack level (Courchamp *et al.* 2000a; see also Section 2.4). Simulations suggest that African wild dogs are likely to disappear once the number of packs falls below a critical value (Fig. 3.11). Ecologically, this is because fewer packs generate fewer dispersers which are in turn less likely to colonize a patch successfully, because they arrive only in small numbers and do not always manage to find other dispersers with which to reproduce. Natural decreases in pack numbers due to pack extinctions are thus not balanced by new colonizations in small metapopulations. In general, the critical number of occupied patches below which the metapopulation as a whole goes extinct is seriously affected by such variables as the Allee effect strength, the emigration rate from patches, migration mortality, the initial size of local populations and the degree of demographic stochasticity (Zhou and Wang 2004).

Allee-like effects of this kind may also occur in populations of parasites spreading through hosts—hosts serve as patches for local populations of parasites. If there is a strong Allee effect in local parasite populations due to, e.g., a need to overcome the host's immune system or find a mate for sexual reproduction (e.g. May 1977a, Garrett and Bowden 2002), too small a number of infected hosts (i.e. too small a number of occupied patches) can cause the infection to

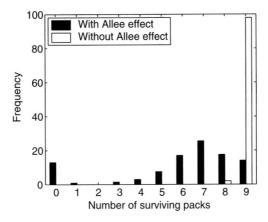

Figure 3.11. Distribution of the number of packs surviving after 100 years in a nine-patch metapopulation model where local populations are either free of any Allee effect (white bars) or are subject to a strong Allee effect (shaded bars); results based on 1000 simulation replicates. Redrawn from Courchamp *et al.* (2000a).

decline rather than spread; see Section 3.6.3 for more on Allee effects, parasitism and disease dynamics.

Allee-like effects can also occur in metapopulations that do not suffer from strong Allee effects in local dynamics. When only a few patches are occupied the number of dispersers is likely to be small; these may not succeed in reaching and hence colonizing suitable empty patches and/or rescuing currently occupied patches from extinction, particularly if local population growth rate is low or dispersal mortality is high, even if local populations grow logistically (Lande *et al.* 1993, Amarasekare 1998a). Since these conditions are apparently more stringent than those given in the previous two paragraphs, metapopulations with Allee effects at the local population level are likely to be more vulnerable to extinction.

Model (3.14) is a coarse, phenomenological description of effects of both these mechanisms. Table 5.1 in Chapter 5 summarizes the main consequences Allee effects might have for local populations as well as the whole metapopulation along continuum of fragmentation, a picture to which models reviewed in this and the previous subsection have contributed a great deal.

3.5.3. **Invasion dynamics**

Invasion is a spatial phenomenon which usually consists of a localized appearance of a small number of plants or animals, establishment of an initial population, and spatial expansion out of its initially small area of occurrence. Allee effects, together with demographic and environmental stochasticity, hamper successful establishment of invaders (or (re)introduced species or biological control agents). Small invading populations have a disproportionately reduced chance of establishment, as shown in Section 3.4 and as demonstrated in a number of comparative studies (e.g. Long 1981, Green 1997, Deredec and Courchamp 2007); see also Sections 5.1.2 and 5.2.4. Invasion dynamics and their interaction with Allee effects has been a topic of wide interest and much work has already been done in this direction (see the comprehensive review by Taylor and Hastings 2005). Table 3.5 summarizes the main consequences that Allee effects might have for invading populations. Some of these model predictions hold equally for strong and weak Allee effects, some are limited only to strong Allee effects.

The models predominantly used to explore the implications of Allee effects for dynamics of invasive species are similar to metapopulation models, but treat space as a continuous entity, although discrete-space models also exist (e.g. Hadjiavgousti and Ichtiaroglou 2004). The former include discrete-time, integro-difference models (e.g. Kot *et al.* 1996, Veit and Lewis 1996, Wang *et al.* 2002) and continuous-time, reaction-diffusion[3] or reaction-diffusion-advection[4] models (e.g. Lewis and Kareiva 1993, Lewis and van den Driessche 1993, Wang and Kot 2001, Petrovskii and Li 2003).

For passive dispersers, dispersal works to dilute the population at any given location, thereby requiring higher initial densities to overcome the Allee effect than in counterpart non-spatial models (Taylor and Hastings 2005). In other words, a founder population subject to a strong Allee effect may fail to establish itself, even when initially at levels which exceed the Allee threshold, because its growth may not be sufficient to offset the decline in local population density through dispersal. On the other hand, population extinction due to high dispersal rates can be preceded by a relatively long transient period of relatively constant population density and distribution (Hadjiavgousti and Ichtiaroglou 2004), a pattern reported for many invading birds (Long 1981). Overall, the success of a founder population in invasion will depend on the initial population density, but also on the shape and size of the area the founder population occupies (Lewis and Kareiva 1993, Kot *et al.* 1996, Soboleva *et al.* 2003). Generally, the larger the initially occupied area, the lower the initial threshold size of local populations for the metapopulation as a whole to grow and expand. Furthermore, the ability of a founder population to grow and expand will depend on the overall habitat size (Table 3.6) and the intensity of advection to which the population is exposed (Petrovskii and Li 2003, Almeida *et al.* 2006).

Once a founder population is established (i.e. grows at a positive rate) it starts to spread geographically on its own. Allee effects can adversely affect rates of

---

[3] No particular direction of population spread: spread by diffusion only.
[4] Population spread by diffusion plus movement in a particular direction, e.g. by ocean currents or prevailing winds.

Table 3.5. Ecological consequences of Allee effects on dynamics of invasive species.

Consequence	Model type	Type of Allee effect	References
Non-spatial consequences			
Species must be introduced at population size or density higher than Allee threshold for invasion to succeed	Deterministic	Strong	Volterra (1938), Odum and Allee (1954) and many others
Probability of establishment declines sharply at Allee threshold	Stochastic	Strong	Dennis (1989, 2002), Liebhold and Bascompte (2003)
Spatial consequences			
Initial population must occupy area larger than critical spatial threshold	Deterministic	Strong	Bradford and Philip (1970), Lewis and Kareiva (1993), Kot *et al.* (1996), Etienne *et al.* (2002), Soboleva *et al.* (2003)
Rate of spread slower	Deterministic or stochastic	Strong or weak	Lewis and Kareiva (1993), Kot *et al.* (1996), Wang and Kot (2001), Wang *et al.* (2002), Hadjiavgousti and Ichtiaroglou (2004), Almeida *et al.* (2006)
Accelerating invasions converted to finite speed invasions	Deterministic, integro-difference with fat-tailed dispersal kernels	Strong or weak	Kot *et al.* (1996), Wang *et al.* (2002)
Range pinning	Deterministic or stochastic with discrete (patchy) space	Strong	Fath (1998), Padrón and Trevisan (2000), Keitt *et al.* (2001), Hadjiavgousti and Ichtiaroglou (2004)
Patchy invasion in continuous habitat	Deterministic or stochastic models with continuous space or lattice models	Strong	Petrovskii *et al.* (2002), Soboleva *et al.* (2003), Hui and Li (2004)
Pulsed range expansion interspersed with periods of relative quiescence	Stochastic model with discrete space	Strong	Johnson *et al.* (2006)

Adapted from Taylor and Hastings (2005).

spread of invading populations. In particular, the rate of spread accelerates only following an initial period of slower expansion, a pattern found in a variety of invading species (Lewis and Kareiva 1993, Kot *et al.* 1996, Veit and Lewis 1996). Using an integro-difference model largely parameterized by data collected on the house finch *Carpodacus mexicanus*, Veit and Lewis (1996) successfully recreated the patterns of temporal changes in the rate of spread of the invading

Table 3.6. Dynamic consequences of the overall habitat size for invasion of passive dispersers (i.e. dispersers spreading via diffusion).

Model type	Logistic growth	Weak Allee effects	Strong Allee effects	References
Non-spatial	Unconditional persistence	Unconditional persistence	Conditional persistence	Chapter 1 of this book
Spatial; infinite habitat	Any initially sparse population will initiate invasion	Any initially sparse population will initiate invasion; actual (asymptotic) rate of spread depends on initial spatial distribution of population density	Dense enough initial populations will spread if Allee effects are not too strong; otherwise population retreats whatever its initial density	Lewis and Kareiva (1993), Hastings (1996), Almeida *et al.* (2006)
Spatial; finite habitat with hostile environment outside	Unconditional extinction for strong diffusion (~small habitat size); unconditional persistence for weak diffusion (~large habitat size)	Unconditional extinction for strong diffusion; unconditional persistence for weak diffusion; conditional persistence for intermediate diffusion (threshold density distribution determining invasion success)	Unconditional extinction for strong diffusion; conditional persistence for weak diffusion	Shi and Shivaji (2006)

The spatial models considered here are reaction-diffusion models without advection.

bird population and in its density near the point of initial invasion—both data and model predictions showed an abrupt acceleration to a constant rate of spatial spread, following an initial period of slower expansion. Models also demonstrate that the (asymptotic) rate of spread is negatively related to the Allee effect strength, including both weak and strong Allee effects (Lewis and Kareiva 1993, Almeida *et al.* 2006).

Sometimes, in theory at least, ecological processes are so finely balanced that 'range pinning' occurs. Range pinning means that even though other suitable habitat is available nearby, the occupied area 'freezes in time' and neither further expands nor shrinks (Padrón and Trevisan 2000, Keitt *et al.* 2001, Hadjiavgousti and Ichtiaroglou 2004, 2006). So far, range pinning has only been demonstrated in models with discrete space in which the range boundary is maintained via a source–sink equilibrium between an inner patch and a neighbouring outer patch.

Invasions can also be considered from the perspective of an individual animal. For example, Allee effects have been shown to reduce the propensity under evolution to leave the currently occupied patch at the invasion front (Travis and Dytham 2002). This is important, since most results on invasion dynamics come from assuming that all individuals disperse passively at a fixed rate. Normally, the risks associated with dispersal vary with direction and/or population density, features that should be more thoroughly considered in future developments of spatial population models.

Geographical spread need not always proceed at a constant rate. In the gypsy moth, a pest in the northeastern United States causing extensive defoliation in deciduous forests, spread takes the form of periodic pulsed range expansions interspersed with periods of relative quiescence (Johnson *et al.* 2006). Modelling shows that a strong Allee effect is a necessary prerequisite for such cyclical invasion dynamics (Johnson *et al.* 2006). A strong Allee effect has indeed been detected in this species, as a consequence of mate-finding difficulties at low population densities (Tcheslavskaia *et al.* 2002, Liebhold and Bascompte 2003, Johnson *et al.* 2006, Tobin *et al.* 2007). This study emphasizes the importance of incorporating Allee effects into models of invasion dynamics if they are suspected to occur, especially if the models are used to assess the efficiency of various management strategies.

3.6. **Allee effects and community dynamics**

Most mathematical models used to study Allee effects are single-species models. But although single-species models have been very useful for understanding Allee effects, their failure to account for interspecific relationships is in many cases an oversimplification. The next step is to include the external ecological drivers of population dynamics of many species—predation, competition, parasitism, mutualism, or any combination of these. Any of these interactions can be affected by Allee effects and demonstrate dynamics which are different from those observed in absence of Allee effects anywhere in the community. We discuss these interactions in turn, considering both methodology and major findings.

3.6.1. **Allee effects and predator–prey dynamics**

We already know that predation (and exploitation) can create component Allee effects in prey (Section 3.2.1). This requires that predator populations do not respond numerically to the target prey species and that the overall mortality rate of prey due to these predators is a hyperbolical function of prey density (equation 3.5). Recall that this situation best fits a generalist predator which is in a dynamic association with another (main) prey, and consumes the target prey as a secondary resource. In this section, by contrast, we are interested in the interaction between the predator and its main prey, assuming that prey, predator, or both suffer from Allee effects which are not due to predation.

The most common framework theoretical ecologists use to explore predator–prey dynamics involves one equation for the prey and one for the predator population:

$$\frac{dN}{dt} = Ng(N) - f(N,P)P$$

$$\frac{dP}{dt} = ef(N,P)P - mP \qquad (3.15)$$

where $g(N)$ is the per capita growth rate of prey, $f(N,P)$ stands for a predator functional response (e.g. as given in Table 3.3, any combination with an aggregative response, or any modification by predator interference or facilitation), m is the density-independent per capita mortality rate of predators and e is a positive constant specifying the efficiency with which energy obtained from consuming prey is transformed into predator offspring. Other variants of this predator–prey model (arguably the simplest) also exist (e.g. Murray 1993, Courchamp *et al.* 2000b, Kent *et al.* 2003, Zhou *et al.* 2005).

Allee effects in prey
Although the literature varies on how to include an Allee effect in prey, the most straightforward way is to use any form listed in Table 3.1 for the per capita population-growth rate of prey $g(N)$. Functions (P2) (e.g. Petrovskii *et al.* 2002, Morozov *et al.* 2004) and (P4) (e.g. Courchamp *et al.* 2000b, Boukal *et al.* 2007) are the two most common examples. An alternative, akin to one of the methods used to modify stock-recruitment models to include Allee effects (Section 3.2.1), is to multiply the term $N\,g(N)$ in prey equation of model (3.15) by $N/(N+\theta)$, where θ determines the Allee effect strength (Zhou *et al.* 2005).

Allee effects in prey generally destabilize predator–prey dynamics. Since the concept of 'population stability' can be viewed from various perspectives (see Box 3.1), so this destabilization can occur on a number of different fronts. Firstly, strong Allee effects in prey can prevent predator–prey systems from exhibiting sustained cycles. Indeed, if troughs in cycles observed in the analogous models without an Allee effect extend below the Allee threshold in prey density, both prey and predators go extinct (Courchamp *et al.* 2000b, Kent *et al.* 2003, Zhou *et al.* 2005, Boukal *et al.* 2007). Secondly, strong Allee effects in prey may destabilize predator–prey systems by causing the coexistence equilibrium to change from stable to unstable or by extending the time needed to reach the stable equilibrium (Zhou *et al.* 2005). Thirdly, predators reduce the equilibrium population size of prey with Allee effects more than that of prey without Allee effects and also enlarge the range of model parameters for which both prey and predators go extinct, thus increasing system vulnerability to collapse (Courchamp *et al.*

2000b). This effect also extends to metapopulations (or rather metacommunities): Allee effects at the local population level generate lower metapopulation sizes and higher risks of metapopulation extinction; Allee effects above a certain strength give rise to inevitable metacommunity extinction (Courchamp *et al.* 2000b). Finally, weak Allee effects in prey cause the predator–prey systems to cycle for a wider range of model parameters than systems without Allee effects, provided that predators have a type II or weakly sigmoidal functional response (Boukal *et al.* 2007).

As Allee effects have been widely demonstrated (or at least surmised) in nature (Chapter 2) and virtually all species are involved in at least one predator–prey (or consumer-resource or host-parasitoid) relationship, we might expect that a number of mechanisms would exist to counteract the negative impact of Allee effects on stability of predator–prey systems. For example, cooperating species (with an Allee effect) often have an increased intrinsic growth rate or are able to reduce predation pressure more effectively than non-cooperators (without an Allee effect); both these processes may compensate for decreased stability due to an Allee effect in prey, e.g. through increased prey population size or density (Courchamp *et al.* 2000b). Also, spatial structuring can stabilize otherwise unstable predator–prey systems (e.g. Taylor 1990, Hawkins *et al.* 1993, Cosner *et al.* 1999 or McCauley *et al.* 2000). Finally, a variety of mechanisms have been suggested that counteract the so-called 'paradox of enrichment' (Boukal *et al.* 2007 and references therein). The paradox of enrichment occurs in models when an increase in the prey carrying capacity (i.e. system enrichment) leads to destabilization of the predator–prey equilibrium and emergence of stable limit cycles (Rosenzweig 1971, Gilpin 1972). The stabilizing property of these mechanisms may likewise counteract the destabilising properties of Allee effects. We note that outcomes of interactions of stabilizing and destabilizing forces are often far from foreseeable and have to be carefully examined in each specific case.

As with single-species models, Allee effects may also alter spatial dynamics of multiple-species interactions. Allee effects in prey significantly increase spatio-temporal complexity of predator–prey systems, especially as regards the ability of both species to co-invade initially empty habitats (Petrovskii *et al.* 2005a). This corresponds to a biological control scenario, where, soon after an alien species invades an empty habitat, its natural enemy is introduced to try and slow down, prevent, or even reverse its geographical spread. There is a fairly wide set of model parameters for which temporal oscillations in both total and local population densities are chaotic while the spatial distribution of both species remains localized and regular; in predator–prey systems with a weak Allee effect in prey or without any Allee effect, the most complicated behaviour is regular population fluctuations (Morozov *et al.* 2004). For some other parameter values, the system demonstrates a patchy invasion, that is, populations spread via formation, interaction and movement of separate patches; another dynamical regime absent when prey are free of any Allee effect (Petrovskii *et al.* 2002, 2005b).

Introducing predators quickly after a successful prey species invasion is key to slowing down or preventing further spread, or even bringing about prey extinction. For example, models of the interaction between the invasive Pacific lupin, *Lupinus lepidus*, and its major herbivore, the moth *Filatima sp.*, on Mount St. Helens show that the chance of reversing the lupin invasion decreases the larger the temporal lag between plant and herbivore arrivals (Fagan *et al.* 2005). The current best estimates of parameters for these two populations place them only slightly above the boundary for lupin invasion collapse, so it will be interesting to see the outcome of future studies in this system (Fagan *et al.* 2005).

The ability of predators to reverse prey invasion (where prey invade in the absence of predators) depends, among other things, on the predator functional response. Prey extinction can be achieved with a lower Allee threshold of prey as the functional response moves from a type II to a type I to a type III; for a type III response, the minimum Allee threshold for prey invasion reversal decreases with increasing steepness of the response (Owen and Lewis 2001). Predators with highly sigmoidal functional responses are thus the best candidates for control of invasion of prey with Allee effects. This may seem counter-intuitive given the relationship between functional response types and predation-driven Allee effects, namely that they are type II functional responses rather than type I and type III functional responses which create an Allee effect in prey (Section 3.2.1). However, here we are dealing with an Allee effect which is not driven by predation. We are also considering spatial structure, and are assuming a dynamic relationship between predators and prey (recall that the lack of predator numerical response is an essential simplifying assumption for the predation-driven Allee effect to occur). The observation that type III functional responses are superior in reversing prey invasion can in part be explained by the fact that in the non-spatial model version, both species are able to coexist in a larger region of model parameters when the functional response is of type III (and of higher steepness) than of type I or type II (Owen and Lewis 2001). The non-spatial coexistence is a necessary condition for the ability of predators to spread spatially and eventual prey extinction; hence the potential for prey invasion reversal increases as the functional response becomes more and more sigmoidal.

Predator–prey, or rather herbivore–plant systems, in which the plant population is subject to a demographic Allee effect, have also been considered from an optimal harvesting perspective. In arid or semi-arid areas where long periods of drought alternate with shorter periods of intense rainfall, a positive feedback (Allee effect due to environmental conditioning) between water infiltration in the soil and vegetation cover has been revealed, generating a critical plant density

(Allee threshold) below which the plant population collapses (Rietkerk and van de Koppel 1997, Rietkerk *et al.* 1997). Overgrazing can obviously push the plant population below this threshold and an optimal policy is therefore needed to ensure long-term sustainability of cattle breeding. A policy whereby a herds-man decreases his stock considerably at the onset of the dry season and returns to a sort of maximum sustainable yield strategy after both plant and herbivore populations recover shortly after the start of a new rainfall season maximizes the long-term income of the herdsman (Stigter and van Langevelde 2004).

Allee effects in predators
In the only study we know of which explored simple predator–prey models with an Allee effect in predators, this Allee effect was modelled by multiplying the term $e f(N,P) P$ in the predator equation of model (3.15) by the term $P/(P+\theta)$, where θ determines the Allee effect strength (Zhou *et al.* 2005). As with strong Allee effects in prey, strong Allee effects affecting the predator population growth rate have the potential to destabilize predator–prey systems by causing the coexistence equilibrium to change from stable to unstable or by extending the time needed to reach the stable equilibrium (Zhou *et al.* 2005). In addition, they increase the equilibrium prey density as compared with predator–prey systems where predators have no Allee effect (Zhou *et al.* 2005).

Emergent Allee effects in predators
It is generally assumed that predators always have a negative effect on their prey, but there is some evidence that predators can also affect prey in a positive way. Positive effects of predators on prey can take on many forms, including the ability of predators to mineralize nutrients which limit prey or prey resources, to 'trans-port' prey to places where intraspecific prey competition is lower (e.g. granivores dispersing seeds), or to alter prey behaviour (Brown *et al.* 2004 and references therein). A model incorporating this kind of positive effect revealed that an Allee effect can emerge in predators via this mechanism (Brown *et al.* 2004), because low densities of predators reduce the availability of their prey.

Positive effects of predators on their prey may also be less direct (and actually not so positive from the prey point of view). In particular, size- or stage-selective predators induce changes in the size or stage distribution of their prey which in turn has a feedback on predator performance. Provided that, in the absence of predators, the prey population is regulated by negative density dependence in development through one of its size or stage classes, and there is overcompen-sation in this regulation such that a decrease in density of the regulating size or stage class will increase its total development rate, predators feeding on a size or stage class other than the regulating one can actually increase the density of the size or stage class on which they feed. As a consequence, an Allee effect emerges in predators—small predator populations cannot invade the prey population since they are not able to induce changes in the prey size or stage distribution which are necessary for their own persistence. The positive relationship between predator-induced mortality and density of the consumed prey class occurs because by eating more, predators reduce competition within the regulating class and cause a higher inflow of individuals into the class they consume (de Roos and Persson 2002, de Roos *et al.* 2003, van Kooten *et al.* 2005). On the contrary, predators feeding on the regulating class will always have a negative effect on their own food density and thus will never exhibit an (emergent) Allee effect (de Roos *et al.* 2003). The emergent Allee effect in predators occurs for a broad range of model parameters (van Kooten *et al.* 2005). Moreover, some parameters may even give rise to a regime in which there are two alternative, stable predator–prey equilibria, similar to the 'predator pit' (van Kooten *et al.* 2005; Fig. 3.12A).

Size- or stage-specific interactions are thus very important to consider, since they qualitatively modify our picture of predator–prey dynamics as compared with when predation is unselective. What's more, the conditions giving rise to the emergent Allee effect in predators appear to occur in many species. In particular, many species of fish, amphibians and zooplankton show clear indications of negative density dependence in maturation rates (de Roos *et al.* 2003); the complex size-structured model in which this kind of emergent Allee effect was initially observed (de Roos and Persson 2002) was parameterized for *Daphnia* (resource), roach *Rutilus rutilus* (size-structured consumer) and perch *Perca fluviatilis* (predator). This emergent Allee effect may occur in many more (predator) populations, and could even be partly responsible for recent collapse and lack of subsequent recovery of many exploited fish populations, including the Atlantic cod *Gadus morhua* (de Roos and Persson 2002). Since anthropogenic effects such as exploitation or habitat degradation impose additional mortality on commercially important or threatened predators, predator populations are more likely to undergo a collapse (extinction or a drop to some small densities; Fig. 3.12A). It has recently been shown that biomass of large predatory fish has been reduced to ~10% of pre-industrial levels (Myers and Worm 2003). This, in turn, imposes shifts in the size or stage distribution of their prey and may affect other species in the community. Restoration of the predator populations will require low predator mortality combined with release of a vast number of predator individuals and/or, somewhat counter-intuitively, removal of the right size or stage classes of prey to attain the prey size or stage structure akin to that in the former predator–prey equilibrium (van Kooten *et al.* 2005; Fig. 3.12B).

Given these studies, a more thorough exploration of positive effects of predators on their prey, both theoretically and empirically, is definitely worth considering in the future.

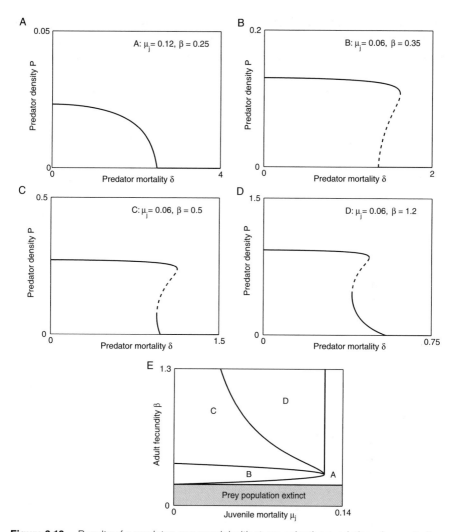

Figure 3.12.   Results of a predator–prey model with stage-selective predation, demonstrating an emergent Allee effect. Panels A to D: Predator density equilibria and their stability (solid: stable, dashed: unstable) as a function of predator mortality rate—(A) No bistability; (B) An emergent Allee effect occurs; (C) An emergent Allee effect together with two alternative stable predator–prey equilibria; (D) Only two alternative stable predator–prey equilibria occur. Bottom: Areas in the two-dimensional parameter space—prey adult fecundity rate β and prey juvenile mortality rate μ_j—in which the regimes shown in panels A to D occur. For sufficiently high μ_j predators can always invade when rare provided that their mortality rate δ is not too high. The shaded area corresponds to model parameters for which prey population cannot persist even in absence of predators. Redrawn from van Kooten *et al.* (2005).

3.6.2. **Allee effects and interspecific competition**

As well as predation, Allee effects modify our view of interspecific competition. All models used to explore this interaction are *ad hoc* extensions of the famous Lotka-Volterra competition model

$$\frac{dN_1}{dt} = r_1 N_1 \left(1 - \frac{N_1 + \alpha N_2}{K_1} \right)$$

$$\frac{dN_2}{dt} = r_2 N_2 \left(1 - \frac{N_2 + \beta N_1}{K_2} \right) \tag{3.16}$$

where r_i is the per capita growth rate of the i-th population, K_i is its carrying capacity, and α and β scale the influence of species 2 on species 1 and vice versa.

Allee effects may affect only one of the competitors and the question is whether an Allee effect will alter the balance of competition between the species. Courchamp *et al.* (2000b) showed that the range of population densities for which a competitor goes extinct is larger if it has an Allee effect. Where both populations coexist, the equilibrium density of the competitor declines if it has an Allee effect as compared with no Allee effect. However, as with prey, many cooperating species that have a socially mediated Allee effect may also have a higher intrinsic growth rate, or may reduce the impact of competitors through cooperation; both these processes may compensate for decreased population levels due to an Allee effect (Courchamp *et al.* 2000b).

The Lotka–Volterra model (3.16) produces three types of dynamic regime, depending on model parameters (e.g. Case 2000): (i) (globally) stable coexistence of both species, (ii) competitive exclusion whereby only one or the other species persists (who wins depends on their initial densities), and (iii) exclusion of one species by the other irrespective of their initial densities. The conditions which lead to these outcomes have been scrutinized in the light of Allee effects present in both species (Wang *et al.* 1999, Ferdy and Molofsky 2002, Zhou *et al.*2004).

Irrespective of model parameters, if both Allee effects are strong, extinction of both species can always occur. Where both species coexist without an Allee effect (Fig. 3.13A), a mild Allee effect (in both species) creates two additional stable equilibria, corresponding to the extinction of one species and persistence of the other (Fig. 3.13B). For stronger Allee effects coexistence ceases to be possible (Fig. 3.13C), and both populations can go extinct even from high densities (Fig. 3.13D). Eventually, for very severe Allee effects, both competitors go extinct irrespective of their initial density. Thus in general, as for predation,

Allee effects destabilize otherwise stable competitive dynamics. In addition, where coexistence occurs, population densities at the coexistence equilibrium decline with increasing Allee effect strength and it takes longer for the system to reach this equilibrium.

For model parameters which lead to competitive exclusion in the absence of an Allee effect (regimes ii and iii above), an analogous sequence of plots can be created, with only plot B absent. The most interesting consequence arises where there is an inferior competitor which is unconditionally excluded without an Allee effect (regime iii); this species may exclude the superior competitor if Allee effects are present (cf. Fig. 3.13E and F).

Figure 3.13B also implies that a species which experiences an Allee effect cannot invade an established population of its competitor unless its density exceeds a critical value. Translated into the metapopulation jargon, this means that migrants of a species are unable to colonize patches where another competitor species has already established. As a consequence, the interaction between Allee effects and interspecific competition creates and stabilizes spatial segregation of the competing species (Ferdy and Molovsky 2002). In other words, where competition would preclude local coexistence, presence of an Allee effect can allow coexistence at a metapopulation scale. Spatial segregation may also prevent an inferior competitor from being displaced by a superior one (Ferdy and Molovsky 2002). Impacts of Allee effects on dynamics of competitive interactions may thus differ when viewed at different spatial scales, and again, we see that spatial structure in a model can play a stabilizing role.

An interesting example of competition occurs in plants where deceptive species (i.e. species not offering any reward to their pollinators) compete for pollinators with rewarding plants. A combination of a behavioural model (pollinators learn that some plants are deceptive) and a population model (dynamics of both deceptive and rewarding plants in a patch) showed that pollinator behaviour induced an Allee effect in deceptive plants. In particular, given a fixed overall density of the plant community (deceptive plus rewarding plants), there is a threshold fraction of deceptive plants in the community below which this plant type goes extinct and above which it approaches a stable fraction within the community (Ferdy *et al.* 1999).

3.6.3. **Allee effects, parasitism, and disease dynamics**

Models of parasite and disease dynamics were first developed by Kermack and McKendrick (1927), 'systematized' by Anderson and May (Anderson and May 1979, May and Anderson 1979), and have since then been elaborated on many diverse fronts. Predictions of disease dynamics as well as explorations of various infection-based pest control and vaccination strategies now depend heavily on

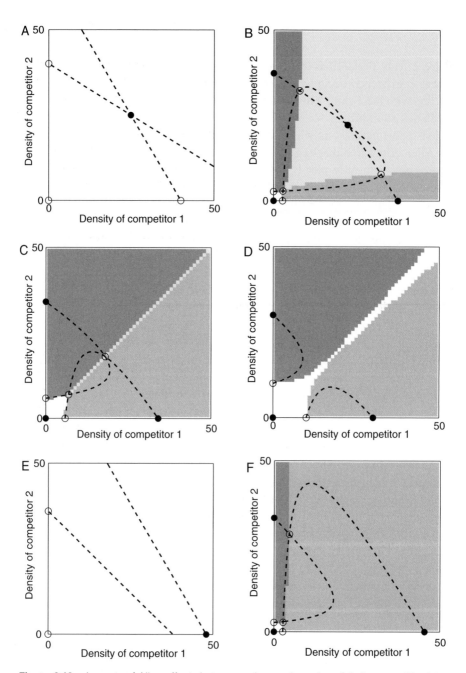

Figure 3.13. Impacts of Allee effects in two species on dynamics of their competitive inter-action. Where the Lotka–Volterra competition model without the Allee effects predicts global coexistence of two competing species (A), two new stable equilibria appear if the Allee effects are mild (B), the coexistence equilibrium ceases to exist for stronger Allee effects (C), and both competitors may go extinct even from high densities if the Allee effects are severe (D). Where the Lotka–Volterra competition model without the Allee effects predicts unconditional exclu-sion of one of the competing species (E), the inferior competitor may persist and exclude the

mathematical models. Parasites are usually categorized either as macroparasites (mainly arthropods and helminths) or microparasites (bacteria, viruses and protozoans, Anderson and May 1979), and models used to predict dynamics of their interaction with their hosts are fundamentally different for these two broad categories. Host-microparasite ('disease') systems are modelled as a set of equations representing dynamics of and flows between different epidemiological stages of the host (such as susceptible, infected or recovered). Host-macroparasite systems are, on the other hand, modelled as a set of equations representing dynamics of and flows between the host and different life-history stages of the macroparasite—macroparasites spend a part of their life-cycle outside their hosts.

The simplest model describing dynamics of a microparasite in a host population is perhaps the so-called SI model

$$\frac{dS}{dt} = b(N)N - m(N)S - T(S,I)$$

$$\frac{dI}{dt} = T(S,I) - m(N)I - \mu I \qquad (3.17)$$

where $b(N)$ is the host's per capita birth rate, $m(N)$ the per capita mortality rate due to reasons other than infection, μ the disease-induced mortality rate, and $T(S,I)$ the transmission rate of the infection between infected (I) and susceptible (S) population members; $N = S + I$ is the overall host population size or density. Two most common forms for the disease transmission rate are the mass action term $T(S,I) = \beta SI$ (often used for air-borne diseases) and the proportionate mixing term $T(S,I) = \beta SI/N$ (commonly used for sexually transmitted or vector-borne diseases), for a positive parameter β, although other forms have also been proposed (Courchamp and Cornell 2000, McCallum *et al.* 2001, Regoes at al. 2002, Deredec and Courchamp 2003). Model (3.17) has seen many modifications (only susceptibles can be allowed to reproduce, pathogens can be transmitted vertically (i.e. between mother and offspring), there may be a latent stage of the disease or a recovery stage from infection, etc.) and has turned out to be a good starting point for exploring interactions between disease dynamics and Allee effects, both in hosts and in parasites.

---

superior one provided the Allee effects are present (F). Dashed lines represent the null clines of the competitors, i.e., population density combinations for which the growth rate of one or the other species is zero, open dots are unstable and full dots are stable equilibria of the system. Different shades of grey (incl. white) delimit areas of attraction of different stable equilibria. The competition model used is described in detail in Wang *et al.* (1999), from which all panels except D are redrawn.

Allee effects in hosts

Allee effects which affect hosts in the absence of any disease are often incorporated in model (3.17) as a hump-shaped form of the per capita birth rate $b(N)$. Strong Allee effects in hosts are of course characterized by an Allee threshold N_A below which the host population inevitably goes extinct. Many infections, on the other hand, face an invasion threshold N_I—a minimum density of susceptibles needed for the infection to spread from an infectious individual (Deredec and Courchamp 2003). The relative value of these two thresholds is important: in epidemiology, $N_A < N_I$ is required for eradicating a disease without simultaneously bringing the population below its Allee threshold; in pest control, conversely, $N_A > N_I$ facilitates pest eradication by an introduced infection (Deredec and Courchamp 2003).

In general, Allee effects in hosts reduce the range of host densities for which an initially rare infection can spread and decrease the disease prevalence where it spreads (Deredec and Courchamp 2006). From this perspective, host populations with Allee effects are better protected or suffer less from the infection than the equivalent host populations with no Allee effect. On the other hand, once the infection is established, the impact of Allee effects on the host population can also be negative, as the infection usually reduces the overall host population density and widens the range of parasite species that can lead the host to extinction (Deredec and Courchamp 2006). Therefore, Allee effects may have opposite effects on host populations threatened by infections: protecting populations against many infections, but increasing the damage if an infection manages to establish.

As with predation, Allee effects have also been shown to affect spatio-temporal patterns of host-parasite dynamics. Consider an invasive species that succeeds in establishing in a local area and starts to spread spatially. A pathogen introduced to control the spread of the host population can slow down or even reverse the invasion, depending on its virulence (Petrovskii *et al.* 2005b, Hilker *et al.* 2005). The reversal occurs when the disease is able to catch up spatially to the expanding host population front. At the edge of this front, when disease-induced mortality is within a certain range, it can both prevent infection from fading out and also outweigh the host population growth (which is low due to the Allee effect and low host density at the range edge). As with predation, the actual spatio-temporal patterns can admit a variety of complex forms (Petrovskii *et al.* 2005b).

Allee effects in parasites

Allee effects may also be inherent in parasites. Many (asexually reproducing) macroparasites and many microparasites need to exceed a threshold in load to overcome any defence of their hosts or to spread effectively in the host population (this kind of threshold behaviour is often referred to as sigmoidal dose-dependent response in parasitology). For example, a beetle *Mulabis pustulata* shows a sigmoidal dose-mortality relationship to a ubiquitous insect pathogenic fungus *Beauveria bassiana* (Devi and Rao 2006) and the per capita transmission rate of the infection of the planktonic crustacean *Daphnia magna* responds sigmoidally to the microparasitic bacterium *Pasteuria ramose* and the fungus *Metschnikowiella biscuspidata* (Ebert *et al.* 2000).

Regoes *et al.* (2002) developed and analysed a model to study the latter relationship, showing that the parasite extinction is always a stable equilibrium. Where the stable host-parasite equilibrium exists, an initial concentration of the parasite has to exceed a critical value (Allee threshold) to invade a host population. This is in sharp contrast to the system in which the infection rate is linear (i.e. the mass action transmission rate)—in that case, the parasite can invade from any initial concentration provided that the host-parasite equilibrium exists (Regoes *et al.* 2002). The need to cross a minimum density was also predicted for the pine sawyer *Monochamus alternatus* providing a reservoir and vector for the spread of the pinewood nematode *Bursaphelenchus xylophilus* in pine forests (provided there is a sufficient density of pine trees); the nematode carries the pine wilt disease and thus supplies the pine sawyer with newly killed trees (Yoshimura *et al.* 1999).

Once entering a host, sexually reproducing macroparasites often need to mate there, and hence may suffer from a mate-finding Allee effect. Examples are found in the sheep tick *Ixodes ricinus* (Rohlf 1969), schistosomiasis and other helminthic infections in humans (May 1977a), or the Karnal bunt pathogen in wheat (Garrett and Bowden 2002). Focusing their model on parasitic nematodes affecting farmed ruminants, Cornell *et al.* (2004) showed that the mean parasite load per host individual varied over the grazing season and depended heavily on both the herd size and the patchiness of the infection (i.e. spatial heterogeneity of parasite larvae in the pasture, which affected the number of larvae ingested simultaneously)—see Table 3.7. In order to invade the host population, there is always a need for the parasite to exceed a threshold density. Above this density, the invasion probability of the parasite grows monotonically with the herd size, eventually approaching 1; below it, there is an optimum herd size (neither small nor large) for the epidemic to occur (Cornell and Isham 2004).

A parasite which has attracted considerable attention from both empiricists and modellers is the bacterium *Wolbachia*. *Wolbachia* lives in the cell cytoplasm of its host (all the major arthropod groups and filarial nematodes), and is transmitted through the egg from mother to offspring. It causes cytoplasmic incompatibility whereby zygotes produced by fusion of an infected sperm and uninfected egg are unviable (Turelli and Hoffmann 1991, 1995, Keeling *et al.* 2003 and references therein). As all other crosses are unaffected, this makes infected females fitter than uninfected ones (the probability that an infected egg produces a viable individual upon an encounter with a random sperm is higher

Table 3.7. Parasite infestation in ruminants: implications of herd size, initial infestation of the pasture and patchiness of the parasitic larvae in the pasture for the per host mean parasite load.

Levels of initial infestation	Non-clumped larval distribution	Clumped larval distribution	
		Weak clumping	Strong clumping
Low	The smaller the herd the higher the worm load per host	The smaller the herd the higher the worm load per host	The smaller the herd the lower the worm load per host
High	No effect of herd size	No effect of herd size	No effect of herd size

There is a mate-finding Allee effect in worms as a female and a male need to be present in the same host to reproduce. Therefore, small, strongly clumped worm populations are most vulnerable to extinction if the ruminant herd is small as well. We emphasize that we are dealing here with the Allee effect in parasites (threshold dose required for infestation) and not in cows—each worm is an individual and each cow is a patch in the metapopulation of cows. Looking at the table from the perspective of cows, however, we might predict an Allee effect in the population of ruminants, too, if we assume (hypothetically) that higher worm load means lower cow fitness. Indeed, there is higher parasite load in small herds in the case of no or weak clumping. See Cornell *et al.* (2004) for more on this study.

than the probability for an uninfected egg) and helps the parasite spread through the host population. Again, modelling shows that in spatially unstructured host populations *Wolbachia* needs to cross two thresholds in order to invade: (i) the transmission efficiency from infected mother to her offspring has to exceed a threshold level (parasite invasion threshold) and (ii) given condition (i), the initial fraction of infected individuals in the host population must exceed another threshold value (Allee threshold) (Turelli and Hoffmann 1991, 1995, Schofield 2002, Keeling *et al.* 2003). This situation resembles the case of deceptive plants competing with rewarding plants for pollinators in the sense that also deceptive plants needed to form a minimum proportion of the community in order to invade (Section 3.6.2). In a spatial context, these thresholds translate into an initial inoculum size and an area the infected individuals initially occupy; the minimum levels of both for the bacterium to invade depend on the details of how infected individuals disperse within the host population (Schofield 2002).

3.6.4. **Allee effects and mutualism**

Two or more species are in a mutualistic interaction if all species involved benefit from the presence of others. Mutualisms can be obligate if a species needs the other for its own (long-term) persistence (e.g. mycorrhizal fungi which can be totally dependent on their plant hosts, while allowing the plants to use the mycelium's large surface area to absorb mineral nutrients from the soil), or facultative when an alternative (but less fit) life history is possible (e.g. ants feeding on the honeydew released by aphids, while protecting them from predation). Obligate mutualisms are less common in nature than facultative ones (Hoeksema and Bruna 2000).

Perhaps the most familiar example of mutualism is the interaction between flowering plants and their pollinators or between fruit-producing plants and seed dispersers. Amarasekare (2004) studied both obligatory and facultative mutualisms in which one of the mutualists was non-mobile (such as a plant) and the other mobile (such as a pollinator or seed disperser), using a metacommunity framework (a set of local communities connected by dispersal). If obligate mutualisms and only one local community are considered, both species go extinct from any initial conditions if the colonization rate of the mobile mutualist is low (Fig. 3.14A), and a strong Allee effect arises if it is high (Fig. 3.14B). For facultative mutualisms, a strong Allee effect arises if the fitness reduction in absence of the mutualist (e.g. fitness reduction in plants due to selfing if pollinators are not present) drops below a critical value (Fig. 3.14C), otherwise both species coexist from any initial conditions (Fig. 3.14D). In a metacommunity composed of two or more local communities, dispersal of the mobile mutualist from source communities can rescue sink communities from extinction and thus maintain regional persistence of both species.

Among obligate mutualists, system bistability arises as a direct consequence of the mutualistic interaction; that is, we have an emergent Allee effect, simply because neither species can live without the other. Among facultative mutualists, however, the Allee effect is not emergent, because the decline in fitness when the second species is absent is incorporated explicitly into the model (Amarasekare 2004).

Sexual reproduction can also be considered a sort of (within-species) obligate mutualism, with many consequences akin to those discussed in the previous paragraphs (see also Section 3.2 for the relationships between sexual reproduction and Allee effects).

3.6.5. **Allee effects, biodiversity, and more complex food webs**

Modelling studies that involve Allee effects and more than two species or more than one type of ecological interaction are still rare, although some of the studies discussed above considered three-species, resource-consumer-top predator food chains (de Roos and Persson 2002, Brown *et al.* 2004). Hall *et al.* (2005) considered the impact of a disease plus a predator on a prey population. The disease was captured by an SI model version (Section 3.6.3) and the predator impacted its prey through its functional response only (i.e. predator population density was assumed constant). The dynamics of this system with no predation

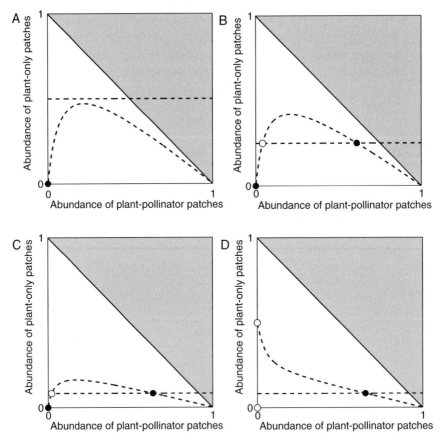

Figure 3.14. Allee effects and mutualistic interactions. Panels A and B: Obligate mutualisms—unconditional extinction of both species for low colonisation rates of the mobile mutualist (pollinator) (A) and bistability (Allee effect) for higher ones (B). Panels C and D: Facultative mutualisms—bistability (Allee effect) when the reduction in recruitment in absence of the mutualist drops below a critical value (C) and the globally stable coexistence in the opposite case (D). The shaded area delimits infeasible combinations of patch abundances—applied models constrain the two patch types to be mutually exclusive and there is an upper bound on the overall number of available patches. Redrawn from Amarasekare (2004) where details of the applied models can also be found.

or a linear functional response were pretty simple—density of susceptible prey had to exceed a threshold value for the parasite to invade its host from any initial level, and the system then attained a (globally) stable host-parasite equilibrium. However, a type II functional response, itself a source of an Allee effect in prey (Section 3.2.1), modified this picture considerably. Depending on the predator density, prey-carrying capacity, and preference of predators for susceptible versus infectious hosts, the system demonstrated a number of dynamic regimes, including Allee effects in hosts/prey, Allee effects in parasites and unconditional extinction of both hosts and parasites (i.e. extinction from any initial density of susceptible and infectious hosts).

Allee effects modify the outcomes of some biodiversity models. The 'neutral theory of biodiversity', appreciated by some and criticized by others, assumes that there are only two major processes that make up a community—speciation and dispersal—and that all individuals within the community are ecologically identical (Hubbell 2001). These simple rules generate species rank-abundance distributions that are remarkably close to some distributions observed in nature, for example in tropical tree communities (although not others, for example coral reefs; Dornelas *et al.* 2006). Imagine a rain forest with a fancy mosaic of canopy trees where tree falls occur from time to time and the gaps are filled by another tree of the same or a different species. In such systems, many species are rare (or appear at low density). This prompted Zhou and Zhang (2006) to ask how the introduction of Allee effects modifies the outcomes of Hubbell's fundamental model. The model works as follows: when an individual is randomly eliminated from a local community, an immigrant is chosen from a metacommunity in proportion to each species' abundance and occupies the opening with probability m. The probability than an individual of species i is recruited locally is $(1-m)\ w_i N_i$ $/ \Sigma_j\ w_j N_j$; the fecundity factor w_i of species i is $w_i = N_i/(N_i+\theta)$, where N_i is the number of individuals of species i in the local community and θ determines the (species-wide) Allee effect strength.

In this rather simple framework, even a relatively weak Allee effect (i.e. with small θ) could decrease species richness considerably. It also resulted in radically different patterns of species rank-abundance distributions than those predicted in its absence, with an excess of both very abundant and very rare species but a shortage of intermediate species (Zhou and Zhang 2006). A good fit between neutral theory predictions and some observations suggests that either Allee effects are not important in such systems or that some stabilizing mechanisms oppose the Allee effect (e.g. Volkov *et al.* 2005).

Allee effects also affect the geographic distribution of species boundaries (Hopf and Hopf 1985). While models of species packing based on negative density dependence in growth predict a stable continuum of species, the model with Allee effects results in a discrete distribution of species along a resource axis. The species separate in a manner that relates to their intrinsic capacities to utilize the resource. Weaker Allee effects slow down the rate of competitive exclusion.

Other multiple-species phenomena remain to be studied in a rigorous modelling framework. The cultivation effect (Section 2.3.2) might be a fruitful area of study, since it involves both competition and predation (Walters and Kitchell 2001). From a broader perspective of large and complex food webs, it seems likely that some species will be exposed to demographic Allee effects and it is possible that any decrease in abundance or density of a species below its Allee threshold may trigger an avalanche of secondary extinctions or at least significant alterations to the wider network of consumer-resource relationships. Adaptivity in foraging decisions and hence a restructuring of the food web might mitigate the impacts of primary extinctions, as might high levels of connectivity within the food web. This again raises fundamental questions about the role of Allee effects in the stability of complex ecological systems. Certainly, much can be learned from exploration of larger food webs in relation to Allee effects, both as regards emergent Allee effects as well as Allee effects explicitly included in models.

3.7. **Allee effects and population stability**

The history of population modelling includes numerous discussions about whether various mechanisms or interactions are stabilizing or destabilizing to population or community dynamics. It is no wonder that Allee effects have also been examined from this perspective.

Allee effects are generally destabilizing to population and community dynamics, and this destabilizing influence manifests itself in a variety of ways. We have already seen, for example, that Allee effects can prevent predator–prey systems from exhibiting sustained cycles, cause the coexistence equilibrium in such systems to change from stable to unstable, extend the time needed to reach the stable equilibrium, or cause the predator–prey systems to cycle for a wider range of model parameters than systems without Allee effects (Section 3.6.1). They can also destabilize otherwise stable competitive dynamics. Where coexistence occurs in the absence of Allee effects, one or both competing species may go extinct in their presence, population densities at the coexistence equilibrium decline with increasing Allee effect strength and it takes longer for the system to reach this equilibrium (Section 3.6.2).

Strong Allee effects make populations more vulnerable to extinction due to the very existence of an Allee threshold. Many population models with Allee effects, including model (3.10), share the property that the Allee threshold and the carrying capacity of the environment are not preset by the modeller, but rather are functions of (some) model parameters. In such models, these two points often approach one another as the Allee effect strength increases (Fig. 3.15A). This is because mechanisms responsible for Allee effect usually reduce the per capita population growth rate over the whole range of population sizes or densities (Fig. 3.15B). If the Allee effect becomes strong enough, the Allee threshold and the carrying capacity may even merge and disappear—no matter how large or dense the population is, it goes extinct (Fig. 3.15A). As a consequence, extinction

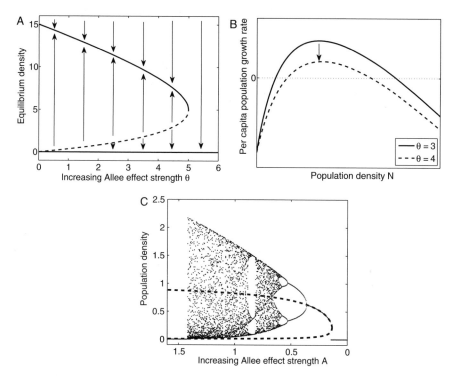

Figure 3.15. Allee effects and population stability. Many population models with a strong Allee effect share the property that changes in the Allee threshold and the carrying capacity implied by changes in a model parameter are negatively correlated (A; arrows indicate how population densities dynamically change in time). This is because an increase in the Allee effect strength (in the arrow direction) results in a decrease in the per capita population growth rate over the whole range of population densities (B). In many discrete-time population models with a strong Allee effect, chaotic behaviour gets stabilised as the Allee effect strength increases (C). On the other hand, decreasing Allee effect strength can cause the population to go extinct (leftmost part of C). Models and parameters used—A and B: $dN/dt = bN\,N/(N+\theta)—mN(1+N/K)$, with $b = 2$, $m = 0.5$, $K = 5$; C: $N_{t+1} = N_t\,exp[r(1-N_t/K)]\,AN_t/(1+AN_t)$, with $r = 4.5$ and $K = 1$. Panel C redrawn from Schreiber (2003).

vulnerability of the population increases with increasing Allee effect strength. Populations with weak Allee effects grow at a lower rate when rare as compared with those without Allee effects, which also makes the former more vulnerable to extinction.

Are there any circumstances under which Allee effects can stabilize rather than destabilize population or community dynamics? We have already seen that Allee effects may affect host populations threatened by infections both positively and negatively: protecting populations against many infections, but increasing the damage if an infection manages to establish (Section 3.6.3). Also, Allee effects can reduce amplitude of otherwise oscillatory dynamics and even eliminate oscillations for some range of parameter values (Fig. 3.15C). In discrete-time population models of the form $N_{t+1} = N_t \, g(N_t)$, including those accounting for Allee effects, an equilibrium corresponding to the carrying capacity becomes unstable once the slope of $g(N)$ at this equilibrium is steep enough (mathematically, if $|1+N \, dg/dN| > 1$ at this equilibrium; see also note to Table 3.1), and regular or chaotic oscillations then emerge in this system in much the same way as in the well-known discrete-time logistic equation (May 1974, Case 2000). Under these circumstances, an increase in the Allee effect strength can stabilize population dynamics, by decreasing the amplitude of oscillations at a higher rate than the increase in the Allee threshold and thus, counterintuitively, reducing the probability of extinction (Fig. 3.15C; Scheuring 1999, Fowler and Ruxton 2002, Schreiber 2003). Once the equilibrium corresponding to the carrying capacity becomes (locally) stable, the system starts to behave as shown in Fig. 3.15A. On the other hand, as the Allee effect strength decreases the population may die out, since the amplitude of chaotic fluctuations grows and may cause population trajectories to fall below the Allee threshold (Fig. 3.15C; Schreiber 2003).

Intuitively, this kind of stabilizing behaviour is not particularly surprising since, qualitatively, an Allee effect acts to reduce the maximum per capita population growth rate which is often positively related to steepness of the $g(N)$ function at the carrying capacity (May 1974, Case 2000). It can also be observed in populations with weak Allee effects or component Allee effects that do not result in demographic Allee effects, but the stronger is the Allee effect, the more powerful is the stabilizing effect (Fowler and Ruxton 2002). For an example of discrete-time Allee effect model which demonstrates chaos but which becomes less stable as the Allee effect strength increases, see Avilés (1999). Arguably, since chaotic dynamics are generally believed to be rather infrequent in nature, if present at all (but see Cushing *et al.* 2003), the potential of Allee effects to stabilize chaos-inducing systems appears to be low. We therefore conclude that Allee effects are more likely to be seen as a mechanism which destabilizes dynamics of real populations.

3.8. Conclusions

Population models have moved on from being regarded by most ecologists as toys for freaky theoreticians to become a vital tool for empirical as well as applied ecologists. They provide formulas for data fitting which help both formalize component Allee effects and assess presence and strength of demographic Allee effects from empirical data. By providing predictions of the outcome of management actions in conservation, exploitation or pest control, they save money where direct experimentation would be costly and allow decision-making where it would be unethical (such as any manipulation of very rare species). Still, however, models serve their theoretical function, by exploring wider population and community level consequences of both component and demographic Allee effects, and thus contribute considerably to building general ecological theory.

In this chapter, we have tried to present the reader with available modelling approaches in a balanced way, focusing on what models used to explore Allee effects look like, what their underlying assumptions are, how and where they can be used, and, most importantly, how they have contributed to our understanding of the impacts of Allee effects on the structure and dynamics of both single-species populations and multiple-species communities. We have considered models in which Allee effects are considered explicitly as building elements, and also models where Allee effects arise as an emergent property. Unfortunately, the limited extent of this chapter does not permit us to discuss many subtleties of structured population models. Also, the models discussed in this chapter focused mainly on population dynamics, with little discussion of evolution as a process which either creates or mitigates Allee effects. It is to this topic that we turn in the next chapter.

In summary, Allee effects have been examined in the context of most existing population model structures, and have significantly altered our picture of population dynamics based on assuming negative density dependence only. Strong demographic Allee effects are of particular importance in this respect because they introduce thresholds in population size or density below which population vulnerability to extinction disproportionately increases. It is important to realize that these thresholds take on different forms in different modelling frameworks. They are single numbers in unstructured population models, higher-dimensional extinction boundaries in state-structured population models, combinations of numbers and spatial distributions in spatially-structured models, and inflection points of the extinction probability as a function of initial population size or density in stochastic models.

The impacts of weak demographic Allee effects, which do not result in negative per capita population growth rates in small or sparse populations, have been rather underexplored in studies of population and community dynamics. This is perhaps because systems with weak Allee effects are generally thought to behave quite analogously to systems with no Allee effects. Hopefully we have convinced you that this is not always the case; recall, for example, that weak Allee effects can prevent successful invasion if the invaded habitat is finite, in much the same way as strong Allee effects in an infinite habitat (Table 3.6). Weak Allee effects may also gain in importance when combined with other (weak or strong) Allee effects, as the resulting double Allee effect can be disproportionately stronger than any of the single ones.

More specifically, we have shown in this chapter, among other things, that:

- there is a great variety of models of component and demographic Allee effects that can be used either alone, as statistical models aimed at data fitting, or as building blocks of more complicated models, to predict broader implications of Allee effects for population and community dynamics;

- Allee effects at a local population level can generate thresholds in the number of occupied patches below which metapopulation vulnerability to extinction disproportionately increases—Allee-like effects;

- Allee effects can significantly modify patterns of invasion as compared with models without Allee effects, such as lowering the rate of geographical spread, accelerating the spread after a period of relative quiescence, or allowing patchy invasion or range pinning;

- Allee effects destabilize predator–prey and competitive dynamics;

- predators or parasites can slow down or even reverse the spread of prey or host populations if the latter are subject to Allee effects;

- Allee effects may have opposite effects on host populations threatened by infections: protecting populations against many infections, but increasing the damage if an infection manages to establish;

- neutral biodiversity models with Allee effects predict radically different patterns of species rank-abundance than those without Allee effects, with an excess of both very abundant and very rare species but a shortage of intermediate species.

This wealth of predictions of population models with demographic Allee effects does not mean that nothing remains to be examined. On the contrary, we expect that future studies will provide us with further insights into many aspects of the interplay of Allee effects and population and community dynamics, especially as regards multiple Allee effects and the role of Allee effects in multiple-species interactions.

4. *Genetics and evolution*

Chapter 2 reviewed the ecological mechanisms which generate component, and in some cases demographic Allee effects. Building on this knowledge, Chapter 3 evaluated the tools available for modelling ecological dynamics of populations with Allee effects, and surveyed the implications of Allee effects for these dynamics. In this chapter, we move on from ecology to consider Allee effects from a genetic and evolutionary perspective.

The topics of Allee effects and of genetics and evolution intersect in four different ways. Firstly, genetics creates its own set of Allee effect mechanisms (e.g. via inbreeding depression). Secondly, there may be genetic differences among members of a population in their susceptibility to ecological Allee effects. Thirdly, we consider Allee effects in the light of evolution—have populations evolved mechanisms to avoid Allee effects, and if so, how is it possible to find anything other than the 'ghosts of Allee effects past' in modern populations? Finally, Allee effects themselves may act as a selection pressure, and members of populations subject to Allee effects may thus evolve different characteristics as compared with those bounded only by negative density dependence. We consider these four issues in turn.

Since the fields of population ecology and population genetics have developed more or less independently, Box 4.1 aims to make the reader's life easier by presenting a glossary of the genetic terms we use in this chapter. Also, because the genetic mechanisms discussed below are only roughly sketched, the interested reader is encouraged to consult any textbook on conservation genetics (e.g. Frankham *et al.* 2002 or Allendorf and Luikart 2006) or any of the works referred to here for more details.

4.1. **Genetic Allee effects**

Small and isolated populations have less genetic diversity than larger or more interconnected ones (Ellstrand and Elam 1993, Frankham 1996, Young *et al.*

Box 4.1.   **A glossary of the genetic terms we use in this chapter**

Allele: An alternative form of a gene. In diploid individuals, each gene is represented by two alleles, one on each chromosome of a chromosome pair; one allele is inherited from the mother and the other from the father.

Allele fixation or simply fixation (of an allele): A situation when (allelic) frequency of an allele in a population equals one. All population members have only this allele at a locus.

Allelic frequency: The proportion of a particular allele in a population at a particular locus.

Allelic richness: Number of different alleles in a population at a particular locus.

Allozyme allelic richness: Number of different alleles in a population at a locus coding for an enzyme.

Census population size: The actual number of individuals in the population (cf. effective population size).

Density-dependent selection: A mode of selection in which the relative fitnesses of different genotypes depend on population density.

Diploid (cell): A cell with two homologous copies of each chromosome, usually one from the mother and one from the father. Each locus carries two copies of a gene (i.e. contains two alleles).

Dominant allele: An allele whose heterozygote phenotype is the same as homozygote phenotype.

Effective population size: Number of individuals in an idealized population that would have the same genetic response to random processes (i.e. lose alleles or become inbred at the same rate) as the actual population of a given size.

Frequency-dependent selection: A mode of selection in which the relative fitnesses of different genotypes depend on genotypic frequencies in the population. It can be negative, in which case relative fitness of a genotype decreases as it becomes common, or positive if the converse relationship is true.

Genetic drift: Random fluctuations in allelic frequency due to chance events in pairing and distribution of parental alleles into a finite number of offspring. Genetic drift is most pronounced in small populations.

Box 4.1. *(Continued)*

Genotype: Genetic makeup of an individual, as distinguished from its physical appearance and attributes (its phenotype).

Genotypic frequency: The proportion of a particular genotype in a population.

Haploid (cell): A cell with only one copy of each chromosome. Each locus carries only one copy of a gene (i.e. one allele). Egg and sperm cells of animals and egg and pollen cells of plants are haploid.

Haplodiploidy: A haplodiploid species is one in which one of the sexes has haploid cells and the other has diploid cells. Most commonly, the male is haploid (develops from unfertilized eggs) and the female is diploid (develops from fertilized eggs: the sperm provides a second set of chromosomes when it fertilizes the egg).

Heterozygote: An individual that carries two (in diploid species) or two or more (in polyploid species) different alleles at a locus.

Homozygote: An individual that carries identical alleles at a locus.

Inbreeding: Fusion of gametes coming from closely related individuals. Inbreeding may also occur through selfing, if it occurs.

Inbreeding depression: Reduction in (a component of) fitness of an individual produced by closely related parents or through selfing (inbreeding) relative to an individual produced by random mating.

Locus (pl. loci): The position of a gene on a chromosome.

Mean population heterozygosity: For one locus, it is the fraction of heterozygotes in a population. If more loci are examined, mean population heterozygosity is an average of heterozygosities calculated for individual loci.

Recessive allele: In diploid organisms, an allele that is expressed phenotypically when present in the homozygote, but that is masked by a dominant allele when present in the heterozygote.

Sampling effect: Reduction in allelic richness in a population as a result of a relatively abrupt collapse. Only a sample of the original set of alleles remains in the resulting small population.

1996, Fischer *et al.* 2000a, Oostermeijer *et al.* 2003). Two major processes are responsible for the decrease (Fig. 4.1): (i) the sampling effect and increased genetic drift, both reducing allelic richness in the population, and (ii) increased inbreeding, reducing mean population heterozygosity. Decreased genetic diversity may

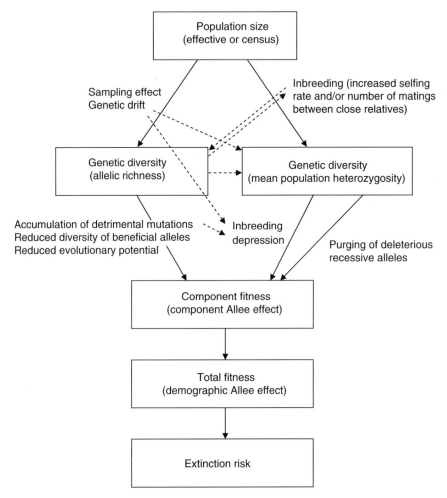

Figure 4.1.   Two major routes (solid arrows) linking population size to individual fitness through genetic mechanisms. Genetic diversity in small populations decreases due to (i) the sampling effect and increased genetic drift, reducing allelic richness in the population, and (ii) increased inbreeding, reducing mean population heterozygosity. Decreased genetic diversity may in turn negatively affect individual fitness, implying that genetic factors can be important in determining population persistence and extinction risk. Purging of deleterious recessive alleles may mitigate the impact of reduced heterozygosity, but seems to be efficient only for alleles which have a large effect. The two routes are not mutually exclusive and there are a number of ways in which one may affect the other (dashed arrows). See the main text for more details.

in turn negatively affect (components of) individual fitness, i.e. individuals in small and isolated populations may suffer reduced (components of) fitness as compared with those in larger or more interconnected ones—we then speak of genetic Allee effects.

Empirical work on genetic Allee effects is still largely suggestive. Despite a number of studies that either demonstrate that the genetic diversity of a population is positively related to its size (e.g. Ouborg *et al.* 1991, Prober and Brown 1994) or that increased genetic diversity of a population brings about an increase in a fitness component of its members (e.g. Byers 1995, Keller and Waller 2002), only a few studies make a full connection between population size and individual fitness, excluding other but genetic causes for their positive relationship and thus revealing a genetic Allee effect. We do not think this scarcity implies that genetic Allee effects are rare, but rather that more work needs to be done to unequivocally demonstrate the importance of genetic effects at a population level. In this section, we first summarize how genetics creates Allee effects, and then present the current evidence for this kind of Allee effects.

4.1.1. Sampling effect and genetic drift

Deficiency in allelic richness can be a direct consequence of a relatively abrupt population decline (e.g. through habitat fragmentation, exploitation, or a natural catastrophe)—we refer to this as the sampling effect, since the removal of a portion of a population means that the remaining individuals contain only a 'sample' of the original allelic pool; in other words, a genetic bottleneck accompanies a bottleneck in population size (Young *et al.* 1996). In populations which are already small and isolated, alleles continue to be lost by chance events; not all parental alleles are represented in a finite number of their offspring. Changes in allelic frequencies arising from these chance events are termed genetic drift. An example of one or both of these effects might be found in the rare, self-incompatible plant *Aster furcatus* in which, except for two rare alleles found in single individuals in three populations, all loci but one of 22 examined were fixed for single alleles (Les *et al.* 1991).

Reduction in allelic richness can affect population viability in several ways. Firstly, in small and isolated populations, mildly deleterious recessive alleles, rather than being eliminated by natural selection, may become fixed through genetic drift. If so, their accumulation in the population will reduce mean individual fitness (Lynch *et al.* 1995). This process is, however, likely to span many generations before there is an observable effect. Secondly, reduction in allelic richness can bring about reduction in diversity of otherwise beneficial alleles. In self-incompatible plants, for example, fixation of the entire population for only one S-allele (responsible for compatibility detection) would have dramatic consequences for plant fertility rate: reproduction would completely stop (Box 4.2). This actually happened in a remnant Illinois population of the Lakeside daisy, *Hymenoxys acaulis var. glabra*; plants in this population did not produce seeds for 15 years because all individuals belonged to a single mating type (all plants shared at least one S-allele; DeMauro 1993). Finally, a decrease in diversity of otherwise neutral alleles can lead to an accompanying decrease in the potential of a species to persist in the face of an environmental change,

---

Box 4.2.   **Self-incompatibility in plants and genetic Allee effects**

Self-incompatibility (SI) is a mechanism that prevents plants from producing seeds following self-fertilization or to some extent also cross-fertilization between individuals sharing a similarity at a particular locus involved in self-recognition (Frankham *et al.* 2002). It is generally thought to evolve to prevent deleterious effects of inbreeding, that is, production of less fit individuals (but see Willi *et al.* 2005 for a counterexample), and involves a recognition step with complex interactions between pollen and stigma that induce acceptance or rejection of the pollen. In particular, SI species have one or more specific loci, the S-loci, typically with tens or even hundreds of alleles, the S-alleles, which control self-incompatibility by disallowing fertilization. There are two basic types of SI (Hedrick 2005). In gametophytic self-incompatibility, the more common type of SI, fertilization results from pollen that has a different S-allele from the female parent. In sporophytic self-incompatibility, the genotype of the pollen parent must differ from that of the female parent; *s* pollen can thus grow on an *ss* stigma if it comes from an *Ss* parent.

   S-alleles can be lost via the sampling effect and genetic drift in small populations, leading to a higher chance of encounter between two incompatible (i.e. closely related) types, reducing per capita seed production and increasing the risk of population extinction. Theoretical models confirm this idea, showing a positive relationship between the effective population size (Section 4.1.4) and the number of S-alleles (Richman and Kohn 1996). Empirical evidence comes, e.g., from the forked aster *Aster furcatus* (Les *et al.* 1991), the Pine Barren thoroughwort *Eupatorium resinosum* (Byers 1995), and the endangered grassland daisy *Rutidosis leptorrhynchoides* (Young *et al.* 2000).

   S-allele diversity is maintained by a strong negative frequency-dependent selection (Frankham *et al.* 2002). As a consequence, if the sampling effect does not eliminate an S-allele, the (effective) population size must be

Box 4.2.   (*Continued*)

rather small to allow genetic drift to overrule the selection and eliminate the S-allele. Byers and Meagher (1992) developed a computer simulation model to show that only populations of less than 50 individuals are unable to maintain a high diversity of S-alleles and suffer from a decrease in the proportion of available mates. In addition, variation among individuals in the proportion of available mates and hence in the per capita seed set is higher in small populations as compared with large populations (Byers and Meagher 1992). Empirical studies of self-incompatible plant populations show comparable numbers: the per capita seed set declines with population size only when the number of plants falls below 10 (Luijten *et al.* 2000), 50 (Fischer *et al.* 2003), or 250 (Widén 1993) reproductive individuals. Genetic Allee effects due to a reduction in the number of different S-alleles can thus be expected to occur only if the (effective) population size becomes very small.

parasites or predators (Ellstrand and Elam 1993). In summary, reduced population size may cause mean individual fitness to decline due to a number of genetic mechanisms—a genetic Allee effect.

By definition, timescales for genetic drift are much larger than for the sampling effect. Nonetheless, the smaller the population, the faster the allelic frequencies will drift and the greater the chance that an allele will be lost from or fixed in the population (Burgman *et al.* 1993). Also, rare alleles are more likely to be lost than more common ones—the probability of fixation of an allele is equal to its initial frequency in the population (Young *et al.* 1996). As a corollary, as more and more loci become fixed for one allele the mean population heterozygosity will decrease.

4.1.2. **Inbreeding**

Small and isolated populations are more likely to experience increased inbreeding than larger or more interconnected ones, due to increased selfing rates in organisms that allow for selfing or to an increased number of matings between close relatives. Increased selfing rates can be a direct consequence of changes in behaviour (higher frequencies of intra-plant pollinations) and/or availability of pollinators—recall the pollen-limitation Allee effect discussed in Section 2.2.1.

Inbreeding reduces mean population heterozygosity relative to a population in which all offspring result from random matings (a naturally outbreeding population). By increasing the frequency of homozygotes in the population, inbreeding allows for expression of deleterious recessive alleles. This results in a decrease in fitness of inbred offspring—a phenomenon termed inbreeding depression—and thus of mean individual fitness in the population as a whole (Charlesworth and Charlesworth 1987, Frankham *et al.* 2002). An increase in homozygote frequency can induce a reduction in mean fitness even if the population is free of deleterious alleles, provided that homozygotes have lower fitness than heterozygotes (heterozygote advantage or overdominance). Finally, inbreeding reduces variability in offspring produced by genetically close parents which in turn decreases the reproductive output (fitness) of the parents in the face of environmental change (Keller and Waller 2002). All these mechanisms are likely to operate simultaneously. Again, since mean individual fitness may decrease with declining population size, a genetic Allee effect may occur.

Inbreeding depression is expected to be most severe just after its onset (that is, just after a population collapse). This is because deleterious recessive alleles which are expressed are then expected to be purged by natural selection, so that after a while mean individual fitness in the population will increase. However, the role of purging in eliminating such alleles and restoring fitness in populations with intense inbreeding still remains a matter of considerable debate (Young *et al.* 1996, Byers and Waller 1999). Purging will be most efficient in eliminating deleterious alleles which have a large effect, generally assumed to be expressed early in the life cycle, but may be relatively inefficient against mildly deleterious alleles which act later in life and are probably much more numerous. These latter alleles will be relatively invisible to natural selection due to their small effect, genetic drift within populations, and gene flow between neighbouring small populations possibly carrying different deleterious alleles (Fischer *et al.* 2000a, Keller and Waller 2002, Willi *et al.* 2005). Therefore, the (effective) population size has to exceed a limit for purging to be at least partially effective (Glémin 2003).

Populations with a long history of regular inbreeding, such as many self-compatible plants or haplodiploid insects, have effectively eliminated many deleterious alleles and thus suffer reduced levels of inbreeding depression relative to predominantly outcrossing species (Husband and Schemske 1996). Indeed, regularly inbreeding species may survive better in small populations than naturally outbreeding species; Jordal *et al.* (2001) showed that numbers of outbreeding species decreased more rapidly with decreasing island size than did those of inbreeders, and that on small islands, outbreeding species tended to be more endemic while inbreeding species were generally more widespread. Somewhat counter-intuitively, inbreeding depression can also decline in a population which becomes fixed for (rather than purged of) some deleterious alleles. This is because in this particular case, the difference in fitness between outbred and inbred offspring is reduced owing to both suffering some 'bad' alleles. Genetic drift can thus lead to a decrease in the extent of inbreeding depression, although the mean individual fitness in the population will in this case remain relatively low regardless of the population size.

4.1.3. **Empirical evidence for genetic Allee effects**

Current knowledge about genetic Allee effects is strongly biased in favour of plants, probably owing to their ease of manipulation. Below we discuss genetic Allee effects in two plant species in detail, leaving other examples for Table 4.1.

Ranunculus reptans is a self-incompatible plant with strong vegetative reproduction in the form of rosettes. Plants originating from large populations had ~29% more flowering rosettes than those from small populations, ~18% more rooted rosettes, and ~23% more rosettes in total (Fischer *et al.* 2000a). Small populations also suffered reduced per capita seed set (Willi *et al.* 2005). Plants in small populations thus showed reduced fitness relative to those coming from large populations. Allelic richness was reduced in small populations, suggesting a limited gene flow and a significant role played by the sampling effect and/or genetic drift in small populations (Fischer *et al.* 2000b). In addition, crosses between plants of long-term small populations (where allozyme allelic richness was used as a proxy for long-term population size) were less likely to be compatible, indicating reduced S-allele diversity (Box 4.2; Willi *et al.* 2005). Offspring from small populations were on average more inbred; inbreeding depression resulted in reduced clonal performance (measured as the number of rooted offspring rosettes produced) and reduced seed production (Willi *et al.* 2005). However, reduced seed production in (long-term) small populations was also observed among outbred offspring; further evidence for the presence of the sampling effect and/or genetic drift (Willi *et al.* 2005). Both the sampling effect and/or genetic drift and inbreeding thus contributed to reduction of individual fitness in small populations of *R. reptans.*

In the rare European gentian *Gentianella germanica,* allelic richness increased with increasing population size and per plant seed set and plant performance increased in populations with higher allelic richness (Fischer and Matthies 1998b). Simply stated, plants in small populations had fewer seeds per fruit and fewer seeds per plant than plants coming from larger populations. The genetic basis of these relationships was confirmed by a common garden experiment; the number of plants surviving to flowering per seed planted, and the number of flowers per seed planted were significantly greater for seeds originating from larger populations than for those from small populations (Fischer and Matthies 1998a). In addition to these (component) genetic Allee effects, small populations of this species suffered a marked decrease in the per capita population growth rate, suggesting a demographic Allee effect: from 1993 to 1995, small populations decreased at a higher per capita rate than larger ones (Fischer and Matthies 1998a, 1998b).

Table 4.1. Some empirical studies demonstrating a genetic Allee effect. Possible mechanisms are the sampling effect, genetic drift or inbreeding.

Species	Genetic Allee effects	References
Royal catchfly *Silene regia*	Reduced germination rates in small populations (germination rates not related to population isolation). Inbreeding depression hypothesized to play a part in the germination rate decrease.	Menges (1991)
Scarlet gilia *Ipomopsis agregata*	Seed mass and germination success significantly reduced in small populations of less than 100 flowering plants. Under simulated herbivory (clipping), plants from small populations suffered higher mortality and grew to a smaller size than plants from large populations.	Heschel and Paige (1995)
Cowslip *Primula veris* and great yellow gentian *Gentiana lutea*	In small populations, plants produced fewer seeds per fruit and per plant. Plants from large populations had a higher seed germination rate, a higher seedlings survival rate, and higher cumulative offspring size per parent plant (mean number of surviving offspring per parent plant times mean offspring size). *P. veris* developed into larger rosettes when seeds were derived from larger populations.	Kéry *et al.* (2000); in *P. veris*, population size was found positively related to allelic richness, an indication of the sampling effect and/or operation of genetic drift (Van Rossum *et al.* 2004)
Ragged robin *Lychnis floscuculi*	Plants from small populations developed fewer flowers than plants from larger populations. Plants from less heterozygous populations developed fewer flowers than plants from more heterozygous populations; this indicates higher inbreeding depression in the former. Correlation between population size and heterozygosity not specified.	Galeuchet *et al.* (2005)
Marsh gentian *Gentiana pneumonanthe*	Offspring from small populations showed reduced fitness (seed and seedling size and survival). Inbreeding depression assumed to play a role.	Oostermeijer *et al.* (2003)
Leopard's bane *Arnica montana* and *Phyteuma spicatum* subsp. *nigrum*	Seed set significantly reduced in small populations. Self-incompatibility and a lack of evidence for reduced pollinator visitation rates suggest that reduced mate availability following a stochastic loss of S-alleles (hence the sampling effect and/or genetic drift) is the underlying mechanism.	Oostermeijer *et al.* (2003)
Small purple pea *Swainsona recta*	Populations of less than 50 sexually reproductive plants were associated with loss of rare alleles and increased inbreeding. Inbreeding associated with reduced values of some fitness components.	Buza *et al.* (2000)
Cochlearia bavarica	Reduced cross-compatibility, reduced fruit set of compatible crosses and lower cumulative fitness (number of plants per maternal ovule) for plants from smaller populations. Both inbreeding and genetic drift likely to play a role.	Fischer *et al.* (2003)

Table 4.1. (*Continued*)

Species	Genetic Allee effects	References
A rare and declining bumblebee *Bombus muscorum*	Allelic richness reduced as compared with a closely related common species. Isolated populations suffer from inbreeding—presence of (sterile) diploid males indicates that the queen mated with a relative.	Darvill *et al.* (2006)
Striped bass *Morone saxatilis*	Reduction in genetic diversity due to population decline (sampling effect and/or genetic drift) suggested to lead to an increased frequency of occurrence of hermaphroditism which in turn inhibits reproduction, further reducing abundance and increasing inbreeding levels in the population.	Waldman *et al.* (1998)
Adder *Vipera berus*	A small and isolated population demonstrated smaller brood size, higher fraction of deformed and stillborn offspring, and lower heterozygosity due to fixation or near-fixation of alleles as compared with larger and non-isolated populations.	Madsen *et al.* (1996)
Florida panther *Felis concolor coryi*	Genetic drift reported to result in kinked tail and poor semen quality in small populations of this species.	Roelke *et al.* (1993)

If we consider that small populations suffer from increased inbreeding (a working hypothesis that needs to be tested in any particular species) the spectrum of species that are likely exposed to genetic Allee effects becomes wide (Keller and Waller 2002). For example, inbred young of many ungulate species suffer higher juvenile mortality than their outbred fellows (Ralls *et al.* 1979). More inbred populations have also been shown to demonstrate higher extinction probability, both theoretically (Mills and Smouse 1994) and empirically (Glanville fritillary butterfly; Frankham and Ralls 1998, Saccheri *et al.* 1998), even when accounting for demographic and environmental stochasticity. This implies that genetic Allee effects, though not necessarily a primary cause of extinction, can result in demographic Allee effects.

4.1.4. **Effective population size**

The concept of population size, central to research on Allee effects, is not an unambiguous one in population genetics. It is the whole recent history of population size, rather than just the actual census size, which determines the genetic structure of populations. This introduces another difficulty into the complex issue of genetic Allee effects.

To standardize the definition of population size, population geneticists developed the concept of the 'effective population size'. In short, the effective population size N_e is the number of individuals in an idealized population that

Figure 4.2. The rare European gentian, *Gentianella germanica,* with fewer seeds per fruit and fewer seeds per plant in small populations, suffers a genetic Allee effect. © Istockphoto.

would have the same genetic response to random processes (i.e. lose alleles or become inbred at the same rate) as an actual population of size N (Box 4.1). N_e is affected by such aspects as individual variance in offspring number, variance in the sex ratio at birth or seasonal fluctuations in population size. A more precise definition of the idealized population and methods to estimate the effective population size under a variety of realistic circumstances can be found in any textbook on population genetics (e.g. Frankham *et al.* 2002).

The effective population sizes N_e is always lower (and can be substantially lower) than the actual census size. Consequently, genetic Allee effects may be observed in populations considered safe at first glance. In plaice *Pleuronectes platessa*, the N_e:N ratio for both the Iceland and North Sea populations was approximately 2 x 10^{-5}, with census sizes estimated at about 10^8 for Iceland and 10^9 for the North Sea (Hoarau *et al.* 2005). According to Hedgecock (1994), this value is not infrequent among marine, broadcast-spawning fish populations, where N_e may be 2–5 orders of magnitude lower than the census size because of the high variance associated with this type of mating system. Recall that broadcast spawning is a mechanism for a non-genetic component Allee effect (Section 2.2.1). Populations of plaice examined after 1970 (following a period of

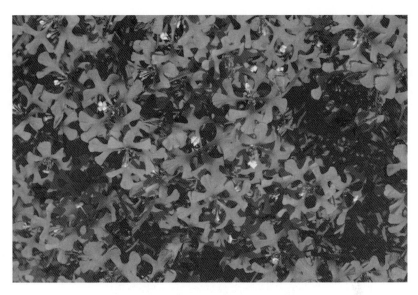

Figure 4.3. Experiments on the annual plant, *Clarkia pulchella*, showed that Allee effects operated on effective population size rather than on census population size.

increased fishing mortality) were characterized by significant heterozygote deficiencies typical of inbreeding (Hoarau *et al.* 2005). Loss of genetic diversity was also recorded in overexploited populations of the New Zealand snapper *Pagrus auratus* (Hauser *et al.* 2002) and the Atlantic cod (Hutchinson *et al.* 2003) in which the effective population sizes N_e were one to two orders of magnitude smaller than in plaice (Hoarau *et al.* 2005). These very small $N_e{:}N$ ratios are not the universal rule, however; estimates of the $N_e{:}N$ ratio across many taxonomic groups, accounting for all the major variables affecting the effective population size, suggest that the average value is around 0.11 (Frankham 1995).

Genetic diversity will have a stronger relationship with the effective population size N_e than with the actual census size N (Ellstrand and Elam 1993). In some studies, genetic diversity was found to be completely independent of the actual population size (Keller and Waller 2002). To explore the role of the effective population size in extinction probability, Newman and Pilson (1997) set up experimental populations of the annual plant *Clarkia pulchella* with the same census size but different effective population sizes. After three generations, germination and survival rates were greatly reduced in small N_e as compared with large N_e populations—mean individual fitness in small N_e populations was only 21% of mean individual fitness in large N_e populations, and 69% of small N_e populations went extinct as compared with 25% of large N_e ones. Genetic Allee effects and apparently also a demographic Allee effect thus operated on the effective population size rather than on the census population size.

Whether N_e or $N_e{:}N$ is a predictor of population vulnerability to genetic Allee effects depends on the history of the population. For example, a population which has a small $N_e{:}N$ ratio because it was smaller in the recent past (in evolutionary terms) than is now is not likely to be vulnerable because it has by definition recovered from small size. Recovery from such bottlenecks is common in pest insects—chemical control measures may push the pest population close to extinction, only for it to recover and outbreak again, often evolving resistance to the chemical. However, recovery following a small $N_e{:}N$ ratio does not rule out presence of a demographic Allee effect since N_e may still have been above a critical value (Allee threshold); consider a population, such as that of *Clarkia pulchella* above, in which N_e rather than N is what drives genetic Allee effects. That said, there is no guarantee that the population will be able to recover repeatedly, given its reduced genetic variation. N_e recovers much more slowly than N, so under a successive series of bottlenecks, we might predict that N_e would continue to decrease and may eventually drop below the Allee threshold, leading to extinction of the population. On the other hand, populations in which a small $N_e{:}N$ ratio occurs naturally (e.g. with mating systems where some individuals contribute disproportionately, where reproduction often fails, or under selective exploitation), would be more vulnerable in terms of genetic Allee effects, as N_e may easily fall below the Allee threshold via increased exploitation mortality or even a mild natural catastrophe. The relationships between the actual population size, effective population size and genetic Allee effects thus require further study.

Overall, there appears to be good reason to suspect that a low N_e might be a more important determinant of a demographic Allee effect than N (and indeed that N may, in some cases, be a misleading predictor of an Allee effect, whether component or demographic). This should act as a warning to conservation managers that seek to conserve highly localized but large populations, particularly if they have historically existed in low numbers (densities).

4.1.5. **Genetic and ecological Allee effects combined**

In a small and isolated population, genetic Allee effects may combine with ecological Allee effects to jointly affect viability and extinction risk. For example, an increased frequency of intra-plant pollinator visits in a small plant population may reduce fitness due to increased selfing rates (genetic Allee effect due to inbreeding). A simultaneous decrease in pollinator visitation rates to the population as a whole may also result in pollen limitation and hence reduce reproductive output of any single plant (ecological Allee effect due to pollen limitation, Section 2.2.1). This interaction may occur in the marsh gentian *Gentiana pneumonanthe*, a rare species in which both inbreeding depression (offspring performance) and pollination success (viable seed production per flower) were found to be significantly and positively related to the number of reproductive individuals in the population (Oostermeijer 2000). A population model built on these observations revealed that while the pollen-limitation Allee effect had little influence on population viability, and increased inbreeding in small populations had a small but significant effect, the strongest reduction in population viability was found when the two component Allee effects acted simultaneously (Oostermeijer 2000) (i.e. the pollen-limitation Allee effect might be dormant in this particular case; see Section 3.2.2).

Similarly, in the Atlantic cod, as in many other broadcast-spawning species (Section 2.2.1, Box 2.6), mean fertilization rate declines with decreasing stock size (an ecological Allee effect). This decline is accompanied by an increase in variance in fertilization rate (Rowe *et al.* 2004), which further leads to a reduction in the effective population size (Nunney 1993, Sugg and Chesser 1994). As a result, genetic Allee effects may intensify the ecological threats which small populations of cod may be exposed to.

Genetic and ecological Allee effects need not always work in concert. Imagine a species that evolves self-incompatibility or some other mechanism to reduce or avoid inbreeding. This will inevitably restrict mate choice for any population member and may as a consequence create or strengthen an ecological mate-finding Allee effect, since mate choice is now restricted to individuals which are sufficiently genetically different. In this context, mate-finding Allee effects and inbreeding are often two sides of the same coin; mate-finding Allee effects are a consequence of the selective pressures imposed by inbreeding (see also Section 4.3.1). As an example, the endangered grassland herb *Gentianella campestris* exists in two plant 'strains', selfing and non-selfing, which thrive under different ecological circumstances; selfing is advantageous in fragmented parts of the grassland habitat facing pollinator deficit, while outcrossing is a fitter strategy in non-fragmented parts (Fig. 4.4; see also Section 2.2.1).

4.2. Demographic Allee effects in genetically structured populations

We saw in Chapter 3 that population models can be structured in various ways, according to individuals' age, developmental stage, body size, sex, space, or even local density. In addition, populations can be structured with respect to individuals' genotype. Not only is then the overall per capita population growth rate a function of the frequency of different genotypes in the population, but also the fitness of different genotypes may be (different) functions of population size or density. We then speak of density-dependent selection.

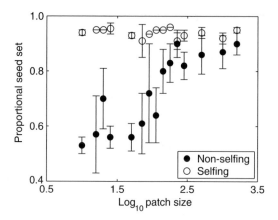

Figure 4.4. Proportional seed set in patches of the field gentian *Gentianella campestris* as a function of patch size for plants which are self-fertile (open symbols) or not (closed symbols). Non-selfing plants show a threshold patch size below which proportional seed set (reproductive output) declines. Redrawn from Lennartsson (2002).

Asmussen (1979) considered a diploid population with one locus and two alleles and hence with three different genotypes. One or more of these genotypes could have density-dependent fitness that obeys a hump-shaped form typical of Allee effects—low fitness at low and high population sizes and a maximum somewhere in between. A strong demographic Allee effect must operate in at least two of the three genotypes for the population-level demographic Allee effect to be strong and hence for deterministic extinction of the population to be possible (Asmussen 1979). Extinction is always possible if low population sizes are detrimental to all three genotypes. When only two genotypes suffer from a strong demographic Allee effect, extinction is possible when either the heterozygote has the lowest fitness at the limit of zero population size or it is the only genotype without a strong demographic Allee effect.

These results imply that if a population with a strong demographic Allee effect is composed of only one allele, a mutation or an inflow of another allele from a neighbouring population could lead to the disappearance of the Allee effect, or at least a change from a strong to a weak one. If Asmussen's major conclusions also apply to situations where there are more than two alleles at a locus (as yet untested), this provides an evolutionary means of avoiding a strong Allee effect, either by mutation or invasion of a 'safe' allele. The results also suggest that Allee effects can be mitigated if the population has the genetic capacity to increase fitness at a given density (including low densities) by increasing the frequency of the genotype with its optimum density closest to the existing density. We are still awaiting an empirical test of these hypotheses.

4.3. **Allee effects in the light of evolution**

One may distinguish two types of rare species: 'naturally rare' and 'anthropo-genically rare'. Many species are naturally rare—i.e. they occur naturally in populations which are small or sparse—and yet do fine. Think of the Devil's Hole pupfish (*Cyprinodon diabolis*) population in Amargosa Valley, Nevada, which fluctuates between 150 and 500 individuals[1] or the rare plant *Kunzea sinclairii* endemic to small rock outcrops and cliffs on Great Barrier Island, New Zealand (de Lange and Norton 2004). On the other hand, many species which were not previously rare, such as the Atlantic cod or rhino, have become so through anthropogenic activities (Chapter 5). Of course, naturally rare species can become anthropogenically even rarer.

This distinction becomes important in the context of evolutionary history. Natural selection works to adapt individuals in a population to the environment in which they finds themselves. The size or density of the population is itself part of that environment. Thus species which are naturally rare have evolved to take advantage of the benefits of being rare, and to reduce as far as possible the associated disadvantages, while species which are naturally abundant have evolved to take advantage of the benefits of being numerous. In other words, species which are naturally rare perform well while being rare, while species which are anthropogenically rare may not. Anthropogenically rare species may thus be more immediately threatened by reduced fitness that may be associated with small population size or low density. Hence, we might hypothesize that large and stable populations will be more susceptible to Allee effects than small or fluctuating ones. Bear in mind, however, that there are many examples of large populations which have suffered a bottleneck but quickly recovered to pre-collapse levels (e.g. Gerber and Hilborn 2001).

A species can be naturally rare for a number of reasons. It may be habitat-specific and thus rare if its habitat is rare (e.g. rock outcrops, cliffs), localized in a geographically restricted area (e.g. small islands), or sparse, i.e. occurring at low densities (e.g. many species of the Proteaceae family from the Cape Floral Kingdom, South Africa.[2] Combining these three categories gives us seven forms of natural rarity and only one form of commonness (geographically widespread, dense populations with no specific habitat requirements). With respect to Allee effects, species that have evolved means to cope with low density may still be susceptible to Allee effects operating on population size (such as Allee effects mediated by anti-predator behaviours), and vice-versa (such as the mate-finding Allee effect).

---

[1] www.fws.gov/desertcomplex/pupfish/devilshole.htm
[2] protea.worldonline.co.za/rabin.htm

For what follows, it is essential to distinguish evolutionary adaptations that operate at low population density (or small size) from those that operate at high density (or large population size). Also, we can distinguish adaptations which only circumvent an Allee effect by avoiding low density and maintaining the population at high density (temporarily or permanently) from those that change the overall Allee effect-related fitness curve (Fig. 4.5). Gregariousness is not an adaptation to avoid an Allee effect—the Allee effect is still there. It is, however, an adaptation to avoid reaching the Allee threshold or to remain at a density which optimizes individual fitness (Fig. 4.5A). Gregariousness can be regarded as (an indicator of) an Allee effect mechanism in the sense that gregarious populations are gregarious precisely because there is some fitness benefit to be gained by avoiding low density. Many of the Allee effect mechanisms discussed in Chapter 2 are exactly of this type. Evolutionary adaptations to mitigate Allee effects, on the other hand,

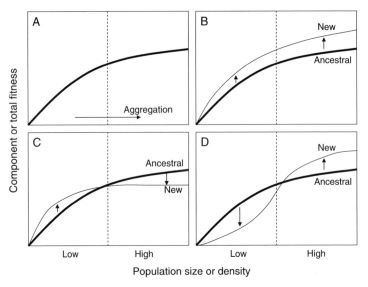

Figure 4.5. Allee effects in the light of evolution. The graphs show the (generalized) relationship between fitness and population size or density for a population with Allee effects; thick line: before evolution to mitigate Allee effects; thin line: after. A. Some evolutionary adaptations operate by enhancing (temporarily or permanently) the local population size or density—the Allee effect is still there but the population finds itself in a 'safer' region. B. Other evolutionary adaptations operate at either low or high density, and raise individual fitness across the whole density range—they mitigate any existing Allee effect. C. Still other adaptations operate at low density and, due to a trade-off, reduce fitness at high density—these also mitigate any existing Allee effect. D. Finally, there are evolutionary adaptations that operate at high density but, due to a trade-off, reduce fitness at low density—these adaptations increase the Allee effect strength and hence extinction vulnerability once the population collapses.

are those which lower or remove the threshold (or more generally, the decline in fitness) and operate at low density. At high density such adaptations may bring about an accompanying fitness increase (Fig. 4.5B) or a fitness decrease (Fig. 4.5C). These are the adaptations we discuss below. Finally, there are adaptations which operate at high density. Again, these may affect low-density populations either positively (Fig. 4.5B) or negatively (Fig. 4.5D); it is the negative impact of such adaptations on low-density populations—via strengthening or appearance of Allee effects—which we discuss at the end of this section.

4.3.1. **Low-density populations—evolution to mitigate or avoid Allee effects**

An Allee effect is a process which affects individual fitness, and is therefore subject to natural selection—in relatively small or sparse populations those individuals which do a little better will have more offspring and pass on more of their genes. Given enough time, members of a rare species should evolve means of mitigating or even avoiding a component Allee effect, just as they evolve means of avoiding predation, starvation and so on. Below, we give possible examples of evolutionary adaptations to mitigate or avoid Allee effects. Do these adaptations represent the 'ghost of Allee effects past'? In other words, should we expect to see no Allee effects, or at best weak Allee effects, in naturally rare populations?

Allee effects might have been responsible for evolution of traits as diverse as mate finding cues (pheromones, songs and calls), dispersal, habitat preferences, mating synchronicity, homing behaviour, gamete morphology and performance, etc…In a modelling study, Jonsson *et al.* (2003) compared the efficiency of a pheromone and a non-pheromone mate-finding strategy typical of two groups of beetles. Whereas there were only small differences between these strategies at high density, the pheromone strategy was more efficient when conspecific density or density of host trees declined.

Beside the fact that we can hardly provide formal proof of the idea that pheromones and other such adaptations arose in response to Allee effects, it is important to note that some of these adaptations need not necessarily be direct evolutionary responses to the presence of Allee effects. For example, advertisement calls of the Cuban tree frog *Osteopilus septentrionalis* apparently help individual males (and females) increase their probability of pairing if females are dispersed and difficult to locate in dense riparian vegetation, but they also serve as a tool in intense intra-sexual competition in male-biased breeding aggregations (Vargas 2006)—indeed, call duration was shown to be an indicator of genetic quality in males of the grey tree frog *Hyla versicolor* (Welch *et al.* 1998). It is possible that the signal had some adaptive function first (mate-finding) before it became exaggerated and used for sexual selection. Of course, this does not rule out the possibility that it originated in large populations to increase the speed at which a female (say) noticed a given male faster than his mute competitors. At the very least, we have to be aware of subtleties and complexities associated with a quest for evolutionary origins of an adaptation.

Mate-finding is a problem that needs to be solved by members of any sexually reproducing species. In a modelling study, Berec *et al.* (2001) showed that an increase in the mate detection distance significantly decreased strength of the mate-finding Allee effect. It is therefore not surprising that most species show some form of adaptation for mate-finding (Mosimann 1958). Traits that have only a negligible adaptive value in large and dense populations might turn out to be vital when populations become small or sparse; one only has to think of the vast variety of methods used to attract a mate across large distances—songs and calls, pheromones, aerial displays etc. Plastic behavioural strategies whereby at low population density the bush cricket moves more (Kindvall *et al.* 1998) or the fraction of calling males of the European field cricket *Gryllus campestris* increases (Hissmann 1990) are other adaptations of this kind. To ensure that pollination can occur over large distances, many sparse plant species have evolved highly specific mutualisms between one plant species and one species of pollinator (often insects but also bats; Sakai 2002, Jackson 2004, Maia and Schlindwein 2006, Muchhala 2006).

Another set of evolutionary adaptations involves those that tend to decrease the frequency at which mate finding has to occur. Such adaptations count a lot at low density, but are possibly of little benefit at high density. They include the ability of females to store viable sperm, as observed in the box turtle (Ewing 1943) as well as in many invertebrate species induced ovulation, such as in the red deer *Cervus elaphus* (Jabbour *et al.* 1994) and ability to maintain long-term or even lifelong pair bonds, as observed, e.g., in the wandering albatross *Diomedea exulans* (Dubois *et al.* 1998). The ability of females to store sperm is important in many marine species, particularly in crustaceans where mating usually has to coincide with moulting and therefore where only short mate-finding windows are available. In the blue crab *Callinectes sapidus* females only mate once, during their terminal moult, and store sperm which is used for the rest of their reproductive life (up to several years); another possible adaptation for mate-finding in this species is mate guarding by males of females getting ready to moult, although male–male competition definitely plays a role here, too (Carver *et al.* 2005).

Such adaptations also involve mating systems as diverse as parthenogenesis, hermaphroditism, or even density-dependent sex determination which all help individuals mitigate or avoid mate-finding problems in low-density populations and hence spread through the population. Parthenogenesis requires no mate-finding at all, and can be permanent, such as in some whiptail lizard species *Cnemidophorus* sp. (Cole 1984), or only seasonal such as in various nematodes,

Figure 4.6. The blue crab, *Callinectes sapidus*, has evolved in a way that mitigates Allee effects: females store sperm for future use when they mate. © Sandy Franz.

rotifers, (parasitic) wasps, mites or aphids. Simultaneous (self-incompatible) hermaphrodites such as hamlets (coral reef fish, *Hypoplectrus* spp.) effectively double the density of mating partners relative to species with separate sexes, since any two individuals can in principle mate. Some hermaphrodites, such as tapeworms and perhaps some molluscs, may even evolve selfing when mate availability is limited (Klomp *et al.* 1964). Finally, copepodites (juvenile stage) of the parasitic copepod *Pachypygus gibber* become males or females depending on availability of sexual partners (and resources) (Becheikh *et al.* 1998).

Rare plants may suffer two opposing Allee effects: inbreeding where self-pollination dominates and pollen limitation where self-pollination is prohibited. Where inbreeding depression is weak, small or low-density populations of outcrossing plants may avoid pollen-limitation Allee effects by evolving self-pollination (Lloyd 1992, Herlihy and Eckert 2002). The field gentian comes in two genotypes—selfing and non-selfing (Lennartsson 2002). Whereas seed set in the non-selfing genotype is strongly related to patch (or plant population) size, with a big reproductive disadvantage to being in a small population, the self-ing genotype has no such disadvantage (Fig. 4.4). This species has recently suffered a lot from fragmentation and loss of its habitat, low-nutrient grasslands, which threatens its further persistence. In the California annual *Clarkia xantiana*, small populations isolated from congeners exhibited reduced herkogamy[3] and protandry[4] (traits which promote avoidance of self-fertilization) as compared

[3] Spatial separation of the anthers and stigma
[4] In this context reproductive asynchrony among sexes whereby males become on average reproductively mature before females

with large populations or small populations with congeners present (Moeller and Geber 2005). Outcrossing is ancestral in *C. xantiana* and the self-pollination seems to have evolved in the small or sparse populations living in inferior, arid habitats and hence suffering lower pollinator visitation rates (Moore and Lewis 1965, Fausto *et al.* 2001). Rare plants may also mitigate pollen-limitation Allee effects by evolving higher pollinator attraction (Haig and Westoby 1988), allocating more into clonal growth (Eckert 2001), or undergoing an evolutionary change from dioecy to simultaneous hermaphroditism (Wilson and Harder 2003)—see also Ashman *et al.* (2004) for a comprehensive review.

Where inbreeding depression strongly reduces individual fitness in small or sparse populations, evolution may allow inbreeding to be avoided. In plants, this may lead to the evolution of self-incompatibility systems controlled by alleles at one or more self-incompatibility loci (S-loci; see Box 4.2), or to the evolution of dioecy[5] or heterostyly.[6] Animals have often been observed to avoid mating with close relatives or disperse before mating, but to what extent these behaviours represent evolutionary adaptations to mitigate or avoid inbreeding remains to be shown. Still, it is possible that these features are more likely to be selected at high density and impose costs at low density once populations are significantly reduced.

Modelling studies confirm these predictions. Cheptou (2004) and Morgan *et al.* (2005) showed that natural selection might lead plants to evolve complete selfing (even when accounting for inbreeding depression) in which case an Allee effect due to pollen limitation disappears. The opposite evolutionary route is also possible, leading plants to become completely non-selfing, with a strong Allee effect due to pollen limitation. Which of these scenarios eventually materializes depends on many population characteristics, including the degree of inbreeding depression, strength of pollen-limitation Allee effect, fertility, and the initial state of partial selfing. It also depends on the self-fertilization mode, with delayed selfing, which gives outcrossing a temporal preference, and prepotency, with outcrossing prevailing when it is possible and selfing prevailing when outcrossing fails, combining advantages of both self- and cross-pollination (Lloyd 1992). Selection for selfing may even lead to population extinction (Cheptou 2004, Morgan *et al.* 2005)—a phenomenon termed 'evolutionary suicide' and discussed further in Section 4.4.1. All these and also many other models of the evolution of self-fertilization in plants consider inbreeding depression to be constant, i.e. neither specified by an underlying model of population genetics nor made at least heuristically dependent on actual population size or density. We expect to gain further insights into this issue once more detailed models are built and analysed. For example, we might find out that the (often) intermediate selfing rates seen in nature could be seen as a trade-off between maximizing

---

[5] Male and female flowers on separate plants of the same species

[6] Plants with different flower morphs characterized by different lengths of pistil and stamens

fertilization on the one hand and avoiding inbreeding depression on the other hand—trying to avoid two kinds of Allee effects simultaneously.

Many of the above ideas and considerations are only suggestive, not conclusive. In our opinion, convincing evidence of evolution to mitigate or avoid Allee effects needs to show that increasing fitness at low density has a cost associated with it at high density, as sketched in Fig. 4.5C. This is not to say that evolution to mitigate or avoid Allee effects must have a cost at high density—the situation drawn in Fig. 4.5B is also plausible—but rather that if we found such an example it would be the most convincing evidence that evolution had occurred to avoid some cost of low density.

Three congeneric sea urchins (broadcast spawners) provide a case study (Levitan 2002a). The urchin (*Strongylocentrotus droebachiensis*) that lives at lowest density has evolved gametes that perform best under sperm limitation (larger eggs and slow, long-lived sperm). On the other hand, the urchin (*S. purpuratus*) that lives at highest density has evolved gametes that perform best under sperm competition (smaller eggs and fast, short-lived sperm). *S. franciscanus* lives at intermediate density and its gametes are somewhere in between. These observations suggest that gametes of *S. purpuratus* would perform suboptimally when density of this species is abruptly reduced, since any mutation of the egg or sperm in the direction of *S. droebachiensis* would result in an advantage that would quickly spread through the population. Similarly, if the density of *S. droebachiensis* were to increase significantly, characteristics of its gametes would be likely to change to resemble those of *S. purpuratus*. Interestingly, *S. droebachiensis,* which has evolved a fertilization system with higher success rates at low sperm concentrations than the other two, bears a cost in the form of higher rates of hybridization with other species at high (total) sperm density (Levitan 2002b). Adaptations that have likely evolved to mitigate reduced fertilization success in low-density populations of broadcast spawning marine invertebrates were widely reviewed by Levitan (1998).

Note also that any example showing that evolution of a trait at high density bears a cost at low density (some examples given below) can be thought of the other way round, namely that adaptations evolving at low density bear a cost at high density, provided that low-density populations still have enough evolutionary potential to reverse such an unsatisfactory state.

4.3.2. High-density populations—the way Allee effects may appear or become stronger

Populations that are naturally large and dense have likely never been exposed to Allee effects. In such populations, natural selection has favoured traits beneficial at large population size or high density; traits that are likely to differ from those preferred in small or sparse populations. Allee effects might therefore appear in such species as a novel feature, or existing Allee effects might become stronger

(Fig. 4.5D), obviously not as a direct consequence of natural selection, but rather as its by-product. They can thus be of great concern once such populations collapse to small numbers or low densities, the more so since they may then also suffer from reduced genetic diversity and hence reduced evolutionary potential to mitigate or avoid Allee effects in turn (Section 4.3.1).

An obvious example is sexual reproduction. Although some controversy still persists, it is generally agreed that the main force which maintains sexual reproduction is the need for continual evolution to escape parasites through a never-ending 'arms race'—a hypothesis commonly known as the Red Queen hypothesis after the Red Queen in Lewis Carroll's Alice Through the Looking Glass, who continually had to run to stay in the same place (Van Valen 1973). Since parasites are more easily transmitted in dense populations, sexual reproduction might have originated in high-density populations. Yet, we have already seen that sexual reproduction can cause Allee effects in low-density populations, through mate finding, pollination, broadcast spawning etc.

Calabrese and Fagan (2004) explored the implications for population dynamics of reproductive asynchrony among individuals, showing that it can decrease the per capita population growth rate at low densities and hence induce a demographic Allee effect. This Allee effect is further exacerbated by protandry. Since reproductive asynchrony and protandry can be favoured by natural selection at higher population densities (Wiklund and Fagerström 1977, Post *et al.* 2001, Satake *et al.* 2001), such evolutionary adaptations could lead to an Allee effect if population density were to become low.

Last but not least, members of a species that became gregarious to circumvent an Allee effect (Fig. 4.5A) may further increase their fitness advantage by behaving 'socially', via group vigilance, defence or foraging, for example. At small population sizes, these adaptations may lead to an increase in fitness; even two individuals may still benefit by cooperating, for example in a hunt (Fig. 4.5B). Alternatively they may lead to increased costs at low density if the behaviour becomes too 'stuck' and lacks flexibility (Fig. 4.5D). The latter case can be exemplified by organisms which have become obligate cooperative breeders (see Section 2.4), such as suricates, African wild dogs, or white-winged choughs *Corcorax melanorhamphos* (Ligon and Burt 2004, Russell 2004), or by myxobacteria where thousands of individuals cooperate to create a fruiting structure from which only a small minority is released in the form of spores so as to establish new colonies and populations (Fiegna and Velicer 2003).

4.4. Evolutionary consequences of Allee effects

A growing number of studies consider Allee effects as a selection pressure on the evolution of various traits, and explore whether populations subject to Allee effects evolve different characteristics as compared with those constrained 'only' by negative density dependence. In this way, almost any ecological issue can be examined for its overlap with Allee effects. Recent (theoretical) applications include the evolution of habitat preferences (Greene and Stamps 2001, Kokko and Sutherland 2001), propensity to disperse (Travis and Dytham 2002), seeding synchrony (Crone *et al.* 2005), and helping behaviour (Lehmann *et al.* 2006). Below, we discuss one application in detail which we find conceptually interesting and of rather general relevance—evolutionary suicide.

Evolutionary suicide is a term commonly used to designate an evolutionary process whereby a trait evolves which has a fitness advantage for the individual which bears it relative to others in the population, but a disastrous effect for the population as a whole, eventually driving it to extinction (Webb 2003, Parvinen 2005, Rankin and López-Sepulcre 2005). A number of population models have been developed to show that evolutionary suicide may occur in theory, and a growing body of experimental and observational evidence suggests that empiricists should also take the concept seriously.

To understand how natural selection may bring a population to extinction, it is important to realize first how evolution and population dynamics interact. The two operate in a loop. For simplicity, consider a monomorphic population (i.e. a population in which all individuals have the same trait, say a body size) at an equilibrium density. A mutation, followed by natural selection, may change the trait (mutants with larger body size invade the population, spread and eventually replace the residents) which in turn may change the balance between births and deaths (larger individuals are better competitors but need to consume resources for longer to produce the same number of offspring), causing the population to equilibrate at a new density (lower density, since birth rate decreases). Ecological timescales being much shorter than evolutionary ones, we assume that the equilibrium density is achieved before a further mutation occurs. This loop runs until an equilibrium trait is achieved in the population for which any further mutant has lower fitness than a member of the population and hence is soon wiped out.

How is this loop related to an increased risk of population extinction as evolution proceeds, and how does it intersect with Allee effects? Consider a population which suffers a strong Allee effect. As a mutant trait invades and replaces a resident trait in the population (an evolutionary step), the extinction risk will increase provided that the carrying capacity and the Allee threshold are brought closer (Fig. 4.7A). After a series of evolutionary steps the extinction risk can become high enough for a stochastic perturbation to cause the population density to drop below the Allee threshold—the population has thus committed evolutionary suicide. As stochasticity plays a decisive role in this sequence of events, we refer to it as the stochastic extinction pathway. Alternatively, the series of evolutionary steps may bring the trait close to a critical value where the carrying capacity and the Allee threshold merge and disappear (Fig. 4.7A). If the equilibrium trait lies above the critical value, the next evolutionary step brings about population extinction—no stochasticity is implied here and we refer to such a sequence of events as the deterministic extinction pathway.

Stochastic and deterministic extinction pathways have their equivalent in models without strong Allee effects. In such a case, the zero population size or density plays a role equivalent to the Allee threshold in models with strong Allee effects. Therefore, unless strong demographic Allee effects are demonstrated in a species, this remains a viable alternative for the hypothesized or observed examples of evolutionary suicide. So Allee effects do not have to be present for evolutionary suicide to occur, but they make it more likely. In addition, if Allee effects are present there is another route through which evolutionary suicide may occur. When the invader replaces the resident, the population as a whole may fall below the Allee threshold and thus go extinct, even if the resident and the invader persist when alone and at sufficiently high densities (Fig. 4.7B). Unfortunately, we currently know of no empirical study that would exemplify this possibility.

Strong Allee effects drove evolutionary suicide in population models which examined the evolution of self-fertilization in plants (Cheptou 2004, Morgan *et al.* 2005), dispersal in metapopulations (Gyllenberg *et al.* 2002, Rousset and Ronce 2004), body size or length of horns in competitive interactions involving contests (Gyllenberg and Parvinen 2001), and searching efficiency of a consumer (Parvinen 2005). The commercially exploited Atlantic cod is sometimes considered a 'classical' example of evolutionary suicide. In this species, preferential fishing of large individuals has induced selection towards earlier maturation and smaller body size (Conover and Munch 2002, Olsen *et al.* 2004). These changes, though preventing the species from quick extinction, have been accompanied by a reduced per capita reproductive output. As the maturation rate increased, the population size gradually declined and the population became more vulnerable to extinction. We put forward some evidence in Chapters 2 and 5 that cod may be subject to a number of component Allee effects, including an Allee effect induced by humans through exploitation. If a potential demographic Allee effect is sufficiently strong (which could also happen even if all of the individual Allee effects are relatively mild—Berec *et al.* 2007; see also Section 3.2.2), evolutionary suicide may occur (stochastic or deterministic) and trigger a decisive cod population collapse.

An example of the last evolutionary step in the deterministic extinction pathway comes from microbiology and concerns the social bacterium *Myxococcus xanthus* (Fiegna and Velicer 2003). These bacteria cooperate to develop into complex fruiting structures during nutrient deprivation, followed by a release of a minority of the population in the form of spores. Experimentally selected

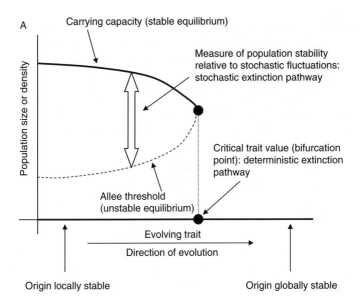

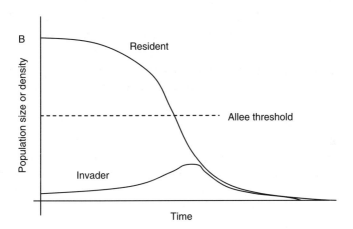

Figure 4.7. Three ways in which Allee effects and natural selection may interact to bring about evolutionary suicide. A. Stochastic extinction pathway: if the carrying capacity of the environment and the Allee threshold approach one another in the course of evolution, gradually lesser stochastic fluctuations may cause the population size or density to fall below the Allee threshold, in which case extinction probability of the population disproportionately increases. Deterministic extinction pathway: alternatively, natural selection may move the population across the critical trait value in which case the population goes extinct. B. Where a resident population is being invaded by a group of fitter individuals and both residents and invaders persist when alone and at sufficiently high densities, both populations go extinct when a strong Allee effect is present. This occurs if an increase in invader density is accompanied by a disproportional decrease in resident density, so that the entire population drops below the Allee threshold.

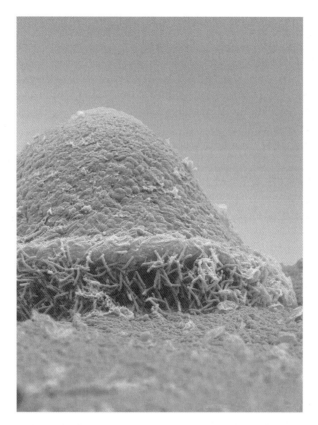

Figure 4.8. The social bacterium, *Myxococcus xanthus,* has evolved unusual cooperative traits © Jürgen Berger and Supriya Kadam/Max-Planck-Institut für Entwicklungsbiologie.

cheater strains, defective at spore production in pure cultures, produce a higher number of spores than the obligately cooperating wild-type strains in mixed cultures, and thus successfully invade and spread through the initially wild-type population. However, as cheaters spread the total spore production and hence the total population density decrease and the extinction probability of the population, including cheaters, increases. *M. xanthus* appears to demonstrate a strong Allee effect—there is a minimum population density necessary to induce fruiting body formation (Kuspa *et al.* 1992).

   As a final remark, we note that traits are likely to never evolve in isolation and that consequences of the evolution of a trait can be overcompensated by the evolution of another one. Regarding evolutionary suicide, this means that in situations which imply a decrease in the per capita reproduction or survival at each evolutionary step, selective pressures in favour of traits that restore reproduction or survival of the population are extremely strong.

4.5. **Conclusions**

Although the concepts of genetic drift and inbreeding are long established in population genetics, and occasionally also mentioned as potential mechanisms for an Allee effect, a more systematic evaluation of genetic Allee effects has begun only recently. A growing body of evidence (mostly from plants) appears to demonstrate that small populations are often genetically 'poor' and that individuals in these genetically 'poor' populations are generally less fit as compared with those from larger populations. Empirical demonstration of genetic Allee effects can be confounded by various factors, including gene flow between neighbouring populations, complexities associated with assessing genetic variance, and the fact that the census population size is not always an appropriate measure of population vulnerability.

One can dream up many adaptations that might have evolved in small or sparse populations as responses to various Allee effects. On the other hand, large or dense populations might have acquired traits that give rise to Allee effects and become increasingly dangerous as the populations get reduced. The amazingly rich world of mating systems and mate-finding adaptations, some of which may have evolved as a response to Allee effects, leads us to speculate about whether there is any chance of observing demographic Allee effects at all, or whether these adaptations, the 'ghosts of Allee effects past', are all we can expect to see in nature. From an evolutionary perspective, we should therefore look for demographic Allee effects in species which are anthropogenically rare, since their history of large and dense populations has not permitted such evolution to occur. Anthropogenically rare species rank high among species which are (or should be) a primary target of our conservation and management efforts, as we shall see in the next chapter.

The strength of genetic Allee effects, and the unambiguousness with which they are demonstrated, as well as the extent to which Allee effects may have contributed to the evolution of a variety of observed individual traits, are only beginning to grasp the attention of researchers. Much remains to be done in this respect. Whereas a general structure for such research seems to be more or less established, we now need to complete the story through well designed experiments linking population size, genetics, fitness and evolution.

5. *Conservation and management*

As the human population rockets towards seven billion, its impact on the global environment is increasingly heavy. For ethical reasons, and also for our own survival, we need to learn how to exploit our resources more intelligently and how to protect our environment from destruction, and the living species it contains from extinction. At this stage of the book, we hope that you are convinced that Allee effects can play an important role in the dynamics of populations which are small, sparse, fragmented, declining, rare, or endangered. Unfortunately, these adjectives apply to more and more populations every year. For those populations which are potentially sensitive to Allee effects, being driven into the small population danger zone may trigger a decline which is difficult to reverse.

This chapter deals with the more applied aspects of the Allee effect; how it can be detected and implications for species which are threatened and for the management of populations, either for conservation or for sustainable exploitation. Because Allee effects act primarily in small and/or sparse populations, they are a vital consideration for the survival of rare, declining, endangered, or fragmented populations. Allee effects are also of major importance in active programs of conservation, such as reintroduction of populations which have gone locally extinct and *ex situ* conservation of highly social populations. They affect how we can manage populations for sustainable exploitation—fishing, hunting, and harvesting—and the impact of less sustainable exploitation—overfishing and poaching. Allee effects can also be very important in the dynamics of species which we don't want—such as non-native species and pest outbreaks.

We start the chapter by placing Allee effects in the context of the issues listed above. We divide this section in two parts; the first dealing with the conservation of rare and endangered populations, and the second with the management of exploited and pest populations. We recognize that there may be overlap between these two categories (exploited species which are also endangered) but nonetheless retain it for the purposes of clarity. We then continue with a review of the methods for evaluating Allee effects in a population, including scientifically rigorous techniques for demonstrating their presence and calculating the Allee threshold, but also more general precautionary-type strategies for assessing the probability of Allee effects being present. At the end of the chapter, we summarize key points as a sort of Allee effect *aide-mémoire* for endangered species and conservation management.

5.1. Allee effects and the conservation of endangered species

There may be a tendency among conservationists to regard Allee effects as a factor which will only kick in when populations are already so small as to be doomed to extinction. Hopefully we have convinced you that this is not the case; recall, for example, that Allee effects can act on both large populations (if sparse) and dense populations (if small). In some cases it may be possible for Allee effects to affect population growth rate and behaviour across the whole spectrum of population size and density.

Thus it is vital that conservation managers engage with the idea of Allee effects, and consider their implications when dealing with small or endangered populations in a variety of different situations. In this section, we consider the impact of Allee effects under habitat loss and fragmentation, in introductions and reintroductions, ex-situ conservation and finally with regard to conservation target-setting.

5.1.1. Habitat loss and fragmentation

Consequences for species with Allee effects
Surviving in habitat that is fragmented and degraded is the most important challenge faced by endangered species on land. Habitat loss obviously reduces overall population or metapopulation size. Habitat degradation may reduce population density, by reducing resource availability (in its widest sense). Both therefore have the potential to interact with Allee effects by bringing the populations closer to any pre-existing Allee threshold. Note here that the key word is 'pre-existing'— habitat loss is not an Allee effect mechanism itself, but it does have the potential to decrease population size or density such that Allee effects can act.

The impact of habitat fragmentation on population dynamics is harder to evaluate. The key issue is the spatial scale of fragmentation relative to the spatial scale at which population operates. The spatial scale of fragmentation has two components, although they are likely to be linked: the (mean) size of individual patches, and the (mean) distance between patches. The size of each individual fragment determines the size of population it can support, and thus the viability of that population (i.e. is it significantly larger than the Allee threshold?). The distance between patches, relative to the dispersal abilities of the species in question, determines population connectivity. The isolation of the fragment and the costs associated with dispersal over intervening habitat determines the likelihood that a population below the threshold will be rescued by immigration from another population within the metapopulation (the 'rescue effect'). Indeed, a high cost of dispersal in a metapopulation may lead to an Allee-like effect at the metapopulation level even if there is no Allee effect at the local population level (Section 3.6.2). Table 5.1 summarizes the consequences of different degrees of fragmentation at different spatial scales for populations with Allee effects.

Reserves as habitat fragments

The potential for Allee effects in fragmented habitat raises issues for the design of protected areas or reserves, since such areas are usually fragments of suitable

Table 5.1. Consequences of habitat fragmentation for populations with demographic Allee effects. We define 'viable population size' as a population significantly larger than the Allee threshold.

Size of fragments relative to viable population size	Isolation of fragments relative to scale of dispersal	Likely consequences for population dynamics	Risk of extinction due to demographic Allee effect
Large	Low	Each fragment can contain several viable populations linked in metapopulation structure	Very low
	High	Viable populations in most fragments, but often isolated	Low—perhaps some loss of populations in small fragments, but even low rates of dispersal may be able to compensate
Medium	Low	Populations in small fragments may go extinct but be rescued by migration ('rescue effect')	Low—high connectivity so small populations likely to be continually rescued by migration
	High	Populations in small fragments may go extinct	Medium—small populations may be lost
Small	Low	Populations depend for survival on connectivity of metapopulation	High in individual fragments; size of overall metapopulation and costs of dispersal key to survival
	High	Populations close to threshold without much possibility of 'rescue' by immigrants	High—small, isolated populations each potentially close to Allee threshold
Very small	Any	Only a few individuals in each fragment	High. Population may not be viable regardless of Allee effects.

habitat surrounded by hostile habitat (farmland, urban sprawl etc.). The optimum design for a reserve system of course depends to a great extent on the geographic area and species requiring protection, but there are general considerations, most notably the so-called SLOSS debate (Single Large Or Several Small reserves), which arose from the theory of island biogeography (MacArthur and Wilson 1967). If habitat islands are close enough together to allow low-cost dispersal between fragments, several small reserves may reduce overall extinction risk by reducing the possibility of a single catastrophic event such as a fire or pollution episode wiping out all the sub-populations simultaneously. They may also be able to incorporate more types of habitat. However, if dispersal is low and/or risky, reserves need to be large enough to accommodate a population well above the threshold size. Models incorporating both stochasticity and an Allee effect show that an intermediate size and number of reserves produces the longest persistence times across a metapopulation (Zhou and Wang 2006).

5.1.2. **Introductions and reintroductions**

The success of introductions and reintroductions hinges on the requirement for a new population to become established, usually from quite a small founding population. A strong Allee effect will create a critical size for the founding population size, and any Allee effect will slow the rate of establishment and thus increase the risk of population loss to stochastic or catastrophic events. Allee effects thus need to be considered when calculating the release size needed for a reasonable chance of success. Furthermore, models suggest that the presence of Allee effects have a qualitative as well as quantitative impact on the most effective release strategy; in the absence of Allee effects, the best strategy is often to have as many releases as possible, to spread risk and stochasticity in time and space. Conversely, in the presence of strong Allee effects, the best strategy might be to have as large a release as possible, even if this means 'putting all eggs in one basket' (Grevstad 1999b).

Many factors influence the success of (re)introductions, including biological factors (habitat quality, genetics, predation, competition, reproduction, age, whether individuals are wild-caught or captive-reared) and non-ecological factors (public relations and education, team management, socio-economics, legislation, time and money available) (Griffith *et al.* 1989, Fischer and Lindenmayer 2000). However, success is generally highly correlated with the size of the initial release (Griffith *et al.* 1989, Hopper and Roush 1993, Berggren 2001). The total number of released animals is more important in determining the probability of success than the number of release events (Wolf *et al.* 1996), and was a consistent predictor of reintroduction success, irrespective of analytical technique (Griffith *et al.* 1989; Wolf *et al.* 1996, 1998). In a review of 180 case studies of reintroductions, Fischer and Lindenmayer (2000) found that releasing more than 100 animals led to higher chance of successful population establishment. Griffith *et al.*

(1989) found that successful translocations had a high mean number of animals released than unsuccessful translocations (160 compared to 54) and suggested a plateau at releases of 80 to 120 individuals for birds and 20 to 40 individuals for large game mammals. Releases of a greater number of individuals significantly increased the establishment success of carnivores (Breitenmoser *et al*. 2001, Fig. 5.1), ungulates (Komers and Curman 2000. Fig. 5.1), diurnal raptors (Cade 2000), birds in Australia and New Zealand (Veltman *et al*. 1996, Green 1997), birds in Europe (cited in Ebenhard 1991) and of a variety of mammals and birds translocated in Pacific countries (Griffith *et al*. 1989) (reviewed in Deredec and Courchamp 2007).

The positive correlation between numbers released and release success may not, of course, be related to Allee effects, but may rather be a simple consequence of demographic or environmental stochasticity. Also, releases of large numbers are often correlated to releases in more sites and over a longer period, so these factors may well contribute (Green 1997). It is also possible that fewer animals are released in situations where the reintroduction is anticipated to fail (Doug Armstrong, pers. comm.). In fact, it is likely that all these effects are important.

Direct evidence of Allee effects acting in reintroduced populations is limited. One exception may be a reintroduced Arabian oryx *(oryx leucoryx)* population in Saudi Arabia, where the authors found evidence of a possible Allee effect when analysing birth rate data from ten years of records (Treydte *et al*. 2001). At low population size (<100 individuals) females showed a low individual birth rate, which increased with population size before declining again in large populations (but changes in rainfall could also account for these results). This Allee effect, if

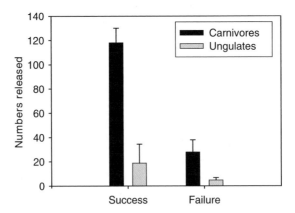

Figure 5.1. Success or failure of reintroductions according to the number of individuals released. For carnivores conclusions on success or failure are taken from the literature, for ungulates from calculating population growth rate (positive = success, negative = failure). From Deredec and Courchamp (2007).

such it was, did not cause the reintroduction to fail since several groups of animals were released over several years, increasing the size of the population above any critical threshold which might exist. This strategy of ongoing 'reinforcement' might be important in helping populations with Allee effects to overcome the initial hurdle of low population growth rate, whether they are reintroduced or simply nearing extinction, at least for relatively long-lived species (Deredec and Courchamp 2007).

A study of Australian reintroductions (Sinclair *et al.* 1998) shows several examples of predator-driven demographic Allee effects in reintroductions. The quokka (*Setonix brachyurus*), black-footed rock wallaby (*Petrogale lateralis*) and brushtail possum (*Trichosurus vulpecula*) all show threshold population sizes below which reintroduced populations had negative growth rates and went extinct (although while a population of 100–150 quokka was in theory stable, population growth rate declined for both smaller and larger populations, suggesting that predation rates are in practice unsustainable). The Eastern-barred bandicoot (*Perameles gunnii*) showed an increasingly negative population growth rate as the population shrinks, although unfortunately in the presence of predators the reintroduced population could never be large enough to achieve a positive population growth rate (Fig. 5.2).

Further examples of introductions are provided by biological control programmes, where considerable research has been done on propagule size for successful establishment of control agents. Exactly the same logic applies to these introductions as to the reintroductions discussed above, although the species concerned (usually insects or pathogens as opposed to vertebrates) have rather different life history characteristics.

Allee effects have been shown to apply in the context of biological control, both to the introduction of the control agent as well as to the hoped-for extirpation of the pest requiring control. Biological control can provide an opportunity for empirical studies of Allee effects, as shown by Grevstad (1999a) who introduced two species of chrysomelid beetles (*Galerucella calmariensis* and *G. pusilla*) into stands of purple loosestrife (*Lythrum salicaria*; their invasive host species) at four different population sizes and monitored changes in population size over three years. For both species, the probability of establishment and the per capita population growth rate increased with increasing initial number of propagules, exactly as for the reintroductions discussed above. In one case, however, a single gravid female was able to found a population which persisted for the three years of the study, so a strong Allee effect is not guaranteed.

Allee effects may account for some of the past failures of biological control: 65% of agents released to control insect pests and 41% of agents released to control weed pests never establish viable populations (Freckleton 2000). An analysis of the outcome of hundreds of parasitoid releases for biological control

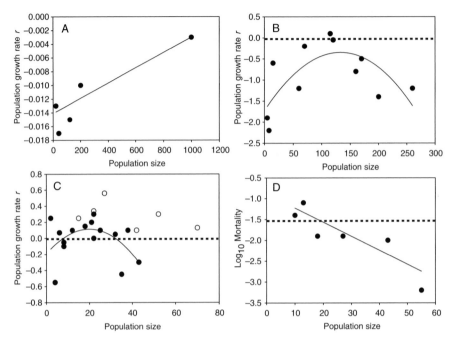

Figure 5.2. Predator-driven Allee effects in reintroductions of four Australian marsupials: A—eastern barred bandicoot (*Perameles gunnii*), B—quokka (*Setonix brachyurus*), C—black-footed rock wallaby (*Petrogale lateralis*), D—brushtail possum (*Trichosurus vulpecula*). For A–C, the y-axis shows population growth rate *r*, with values >0 indicating population increase and values <0 population decrease; 0 is indicated by the dotted line where necessary. Significant fits are shown by a solid line. For D the y-axis shows the mortality rate, with the mortality rate giving *r*=0 indicated by the dotted line (Sinclair *et al.* 1998).

found that the probability of establishment of three taxa (chalcidoids, ichneumonoids, tachnids) depended strongly on the number of individuals released (Hopper and Roush 1993). The study concludes that unless environmental stochasticity is high, the best strategy where a limited number of animals is available is to release all of them at the same time and place to mitigate the impact of Allee effects (and/or demographic stochasticity) as far as possible. Of course, this is a risky strategy since a catastrophic event could wipe out the entire population regardless of density. Also, some parasitoid species may not be affected by Allee effects (Fauvergue *et al.*). A more sophisticated approach might be a mixed strategy, with managers hedging their bets between the security of multiple sites and the higher success probability for larger initial population sizes, or by a long-term programme of release followed by reinforcement (Deredec and Courchamp 2007). Sophisticated modelling techniques are now available to find optimum strategies for a given situation (Grevstad 1999b, Shea and Possingham 2000).

5.1.3. **Ex-situ conservation**

Ex-situ conservation (generally via maintaining self-sustaining populations in zoos) is another situation which by necessity involves small populations. However, captive populations are never limited by resources (shelter and food), predators or mate-finding, so most Allee effect mechanisms can be excluded immediately. However, Allee effects may arise in captive populations via social and behavioural facilitation of reproduction.

In Section 2.2.4 we discuss the idea of 'reproductive facilitation', which arises where individuals need the presence of conspecifics to come into physiological condition to reproduce. For example, pairs of mongoose and black lemur (*Eulemur mongoz* and *E. macaco*) reproduce more readily when conspecifics are put in neighbouring cages (Hearn *et al.* 1996). Larger captive flocks of flamingoes (*Phoenicopterus ruber*) also seem to have better reproductive success because of stimulation into reproductive condition by exposure to reproductive displays by other flock members, male and female (Stevens and Pickett 1994, Studer-Thiersch 2000).

Another potential mechanism in small captive populations relates to female mate choice. A wide number of studies on mammals and birds in captive, reintroduced and endangered populations suggest that females have higher reproductive success given a large pool of males from which to select a mate. When females are given no opportunity to choose (presented with only one male), they may fail to mate or suffer lower reproductive success (e.g. in captive populations of gorillas, cheetahs, leopards, pandas, and kangaroo rats; see review in Moeller and Legendre 2001). Captive groups can also suffer from inbreeding depression (e.g. wolves; Laikre and Ryman 1991). Overall, however, a brief survey of the literature suggests that Allee effects in captive populations are probably a less important cause of reproductive failure than social and behavioural factors arising out of unnaturally close proximity to conspecifics, the artificial environment, behavioural constraints, anomalous behaviour in captive bred individuals etc.

5.1.4. **Target setting for conservation**

The management of a rare or endangered species often includes setting a numerical target: either a size below which the population should not fall, or a target size to be reached. There are a wide variety of goals and techniques used in setting these targets, including genetic sustainability, demographic sustainability, ecosystem function, historical baselines, sustainable harvest etc. (for review see Sanderson 2006).

Where targets are based on previous knowledge about viable populations, Allee effects are already implicitly taken into account, in the sense that such populations must have been above any Allee threshold. However, such data is usually not available, or else the targets they generate are, while desirable, rather unrealistic (Jackson *et al.* 2001). More generally, targets are set by using a combination of models and data to assess a 'minimum viable population' required to meet the target, whether that be to avoid inbreeding, reduce extinction probability or maintain the role of that species in the ecosystem (Sanderson 2006). Whatever the conservation goal, Allee effects need to be taken into account explicitly with this type of target setting.

Maintaining evolutionary potential

Populations need to retain genetic variability in order to maintain evolutionary potential. This means avoiding the loss of alleles and minimizing inbreeding and genetic drift (see Chapter 4). Various rules of thumb have been proposed to achieve this: one migrant per generation or less than 1% inbreeding are two examples (Franklin 1980, Mills and Allendorf 1996). The latter requires an effective population size (N_e) of 50 since the amount of inbreeding per generation (roughly defined) is calculated as $1/(2 N_e)$ (Frankham *et al.* 2002), while conserving genetic diversity in the long term may require an N_e closer to 500 (Franklin 1980).

These thresholds may or may not be adequate in individual populations, but generally, it seems unwise to base conservation policy on only one aspect of the ecology of the target species. Genetic Allee effects are likely to arise in most species at small enough population size, but other Allee effects may also act. For example, in a declining plant genetic and non-genetic Allee effects may act together to give a much higher critical threshold than each would give individually (Oostermeier 2000).

Population viability analysis

Population viability analysis (PVA) is essentially any process which tries to make predictions about the probability of a population either going extinct or persisting over a given time scale (Boyce 1992). It tries to answer questions such as 'What is the probability that my orchid population will go extinct within twenty years?' or 'How many elephants do I need in my park to give me a 95% probability that the population remains viable for 100 years?' Generally speaking, answers to these questions are supplied by population models, which can be simple projections of population time series, or complex models taking into account anything from social behaviour to the spatial arrangement of habitat to the genetic structure of the population (a nice set of examples are models for northern spotted owl conservation; see for example Lamberson *et al.* 1992, McKelvey *et al.* 1993). Modern PVA models seem to be fairly successful at predicting population trajectories over one or a few decades, as long as good data is available on which to base the model, and stochastic catastrophic events such as droughts or fires are not a common feature of the system (Brook *et al.* 2000, Coulson *et al.* 2001).

In Chapter 3 we discussed methods of incorporating Allee effects into population models, so here we will just mention a few key points in regard to Allee effects and PVA.

- *If Allee effects are suspected, they must be incorporated.* If Allee effects are present and not included in the PVA, the predictions will be over-optimistic about the extinction probability of the population. One way round this is to incorporate into the model a 'quasi-extinction' threshold set at some chosen population size. Clearly this needs to be informed by some understanding of the species' ecology, and there needs to be (if possible) good reason for supposing that this point coincides with, or preferably is greater than, the Allee threshold (Ginzburg *et al.* 1982, Liermann and Hilborn 2001).

- *Estimates of stochasticity are crucial.* A deterministic model of Allee effects will produce an extinction probability of zero as long as the population remains above the threshold. It is unlikely that any population analysis nowadays would be so naïve, but a model which underestimates stochasticity will also underestimate extinction risk due to Allee effects at population sizes above the Allee threshold (Dennis 2002, see Section 3.9). More generally, simplified models, which can perform well in estimating means, frequently underestimate uncertainty. In a population where critical thresholds are present, uncertainty can lead to disaster. Estimates of uncertainty and variability in parameter values also need to be quantified realistically (Coulson *et al.* 2001). Tools such as meta-analysis can be useful for putting a realistic range on parameter values; see Hilborn and Liermann 1998 for an introduction.

- *Rules of thumb can be dangerous.* Various rules of thumb are sometimes quoted in relation to minimum viable population sizes for vertebrates, aside from those quoted above with regard to maintaining genetic diversity. It has been suggested that vertebrate populations above ~200 are secure in the short to medium term (Boyce 1992), while other estimates are as high as 7000 (Reed *et al.* 2003). Clearly these sorts of generalizations, particularly at the lower end of the range, should be treated with caution if Allee effects are suspected. For instance, three of the largest remaining populations of African wild dog number ~950 (Kruger National Park, South Africa), 700–850 (Northern Botswana) and ~365 (Selous Game Reserve, Tanzania). All three populations are more or less stable, with the Botswana and Tanzania populations predicted to remain so, and the South African population to contract (Creel *et al.* 2004). Their effective population size will be smaller, but in any case, is not likely to be all that relevant, since demographic and ecological rather than genetic effects are driving the African wild dog to extinction.

Their actual population sizes are higher than some of the 'rule of thumb' thresholds proposed above, but they do not appear to be secure in the long term, and the largest population is in this case apparently the least secure. Conversely, a threshold of 7000 may be over-cautious, even in a species such as the African wild dog with well-documented Allee effects (see Section 2.4). More generally, given that Allee effects are (i) probably widespread and (ii) highly variable in both mechanism and demographic consequences, the idea that some numerical value, whether 50 or 7000, could apply across a wide range of species seems unlikely (Brook *et al.* 2006).

Ecological targets

Conservation targets for some populations may take into account the role of the species in the wider ecosystem; populations which are still extant can nonetheless be regarded as 'ecologically extinct' (Estes *et al.* 1989). This may occur if the population is orders of magnitude smaller than the 'natural' level (see Jackson *et al.* 2001 for various examples) or if the range of the population has contracted significantly. In this latter case, Allee effects may come into play when considering the conservation of populations at the edge of the range, and will affect the ability of a population to expand into areas from which it has been extirpated, or to survive reintroduction into these areas (see above).

IUCN Red List

In creating its 'Red List' of threatened and endangered species, a vital tool for conservation awareness, IUCN is obliged to set explicit population 'targets' in reverse—values below which a species increases its categorization from vulnerable to endangered to critically endangered.[1] These thresholds are necessarily highly general, and therefore, of course, open to criticism. However, IUCN do endeavour to take Allee effects at least partially into account, not directly through the definitions of the categories themselves, but via their definitions of population size. Specifically, IUCN states that in a census of mature individuals, 'Mature individuals that will never produce new recruits should not be counted (e.g. densities are too low for fertilization)'. It is not clear, however, exactly how this operates in practice. For example, in a social species where mature offspring remain in the group as 'helpers', their ability eventually to disperse and form new groups may be severely constrained by Allee effects (see Section 2.4), but it is impossible to assess during a population census whether they will 'never' produce new recruits; the knowledge of ecologists and managers in interpreting population data, with Allee effects in mind, is thus vital.

[1] See www.iucnredlist.org/info/categories_criteria2001

5.2. **Allee effects and population management**

5.2.1. **Exploitation of populations with Allee effects**

Exploitation can have multiple impacts on populations and ecosystems, including changes in food web structure, trophic cascades, and habitat modification (e.g. Estes *et al*. 1998, Jennings and Kaiser 1998, Jackson *et al*. 2001). However, leaving these complications aside, the most straightforward, and usually most important impact of exploitation is an increase in mortality—usually adult mortality—in the exploited population. This increase can be very dramatic (Fig. 5.3).

Increases in mortality interact with the Allee threshold in several different ways. Firstly, an increase in adult mortality is likely to reduce the population size or density, thus pushing the population closer to the threshold. Secondly, increased mortality increases the threshold (Dennis 1989, Stephens and Sutherland 1999, Gascoigne and Lipcius 2004b, 2005). Finally, exploitation can turn a weak Allee effect into a strong one, creating an extinction threshold where previously there was

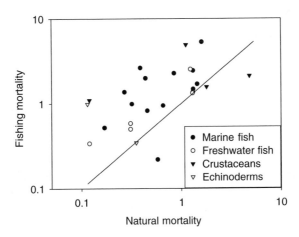

Figure 5.3. Published estimates of natural and fishing mortality for a variety of exploited populations with a broad geographical and taxonomic spread (units year^{-1}), with 1:1 line for reference (representing a doubling of total mortality by fishing). On average, total mortality has increased by ~a factor of 3 (median 2.9, mean 3.9). However, attempts to estimate mortality may reflect a perception that the population is at risk—i.e. there may be a bias in favour of heavily fished populations. Data from Fu *et al*. (2001), Martinez-Munoz and Ortega-Salas (2001), Adam *et al*. (2003), Chen *et al*. (2005), Zhu and Qiu (2005), Kamukuru *et al*. (2005), Al-Husaini *et al*. (2002), Ama-Abasi *et al*. (2004), Sivashanthini and Khan (2004), Olaya-Nieto and Appeldoorn (2004), Mehanna and El-Ganainy (2003), Allam (2003), Hightower *et al*. (2001), Waters *et al*. (2005), Weyl *et al*. (2005), Jutagate *et al*. (2003), Ofori-Danson *et al*. (2002), Langeland and Pedersen (2000), Xiao and McShane (2000), Tzeng *et al*. (2005), Frusher and Hoenig (2001), (2003), Lee and Hsu (2003), Reyes-Bonilla and Herrero-Perezrul (2003), and Morgan *et al*. (2000). Estimates for within marine reserves not included. If estimates separated by sex only females included. Where range given median value is used.

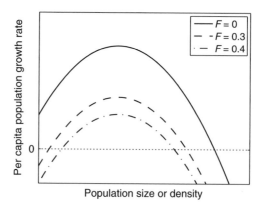

Figure 5.4. Theoretical impacts of fishing (or exploitation) mortality (F) on a population with a demographic Allee effect. A critical Allee threshold is present where the curve crosses the zero line on the left. Exploitation may change a weak Allee effect into a strong one (transition from $F=0$ to $F=0.3$). Where there is a strong Allee effect, increasing exploitation increases the corresponding Allee threshold (transition from $F=0.3$ to $F=0.4$).

none (Fig. 5.4). Attempting to exploit the population for the 'maximum sustainable yield' is particularly hazardous where there is a demographic Allee effect (Dennis 1989, Stephens *et al.* 2002b; Fig. 5.5). Fortunately this type of management has fallen out of favour in recent years, and is now used, if at all, in a much more ecologically realistic way (Hilborn and Walters 1992, Punt and Smith 2001).

Collapse and recovery of fish populations
The majority of examples of heavily exploited populations come from fisheries, and fisheries scientists have long been aware of the possibility of Allee effects (e.g. Myers *et al.* 1995, Liermann and Hilborn 1997 among many other examples). However, research on Allee effects in fisheries has been conducted in a parallel scientific universe to research on terrestrial population dynamics; fisheries scientists even have a different term for Allee effects: depensation.[2]

Fisheries scientists have had a big incentive to investigate 'depensatory processes' (Allee effects), because many stocks have been exploited to low density (an analysis of 232 populations of commercially fished species showed a median decline in population size of 83%; Hutchings and Reynolds 2004). Some of these populations have not behaved according to standard models (which assume logistic-type growth and hence high fitness at low density), the classic example being North Atlantic cod from the Canadian Grand Banks (see Box 2.6).

[2] The emphasis of fisheries studies has been slightly different, with more focus on population level processes and less on individual fitness and behaviour; by necessity since very little is known about individual behaviour in most fish.

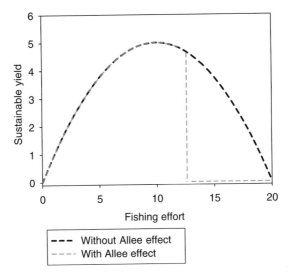

Figure 5.5. The simple fisheries model of equilibrium sustainable yield as a function of fishing effort, with the maximum sustainable yield achieved at intermediate levels of biomass and fishing effort (described by Hilborn and Walters (1992) as '*perhaps the most commonly printed illustration in fisheries textbooks, and the most dangerous*'.). The black line is a model with no Allee effect. The grey line represents a model with a strong demographic Allee effect added. Managing the stock to try and achieve maximum sustainable yield becomes highly risky under these circumstances as stochastic fluctuations can easily bring populations below the Allee threshold (Dennis 1989).

This cod population is not the only heavily fished population which has failed to recover. Hutchings (2000, 2001) analysed population time series for 36 populations of marine fish which showed a minimum 15-year decline in population size, followed by some release from fishing mortality (not generally to zero unfortunately). With the exception of the clupeid family (herrings and relatives, known for their naturally highly variable populations) most populations had recovered very little even with reduced fishing effort (Fig. 5.6). Overall, recovery rates depended on the proportional decline in population size, with populations which had declined by less than half recovering more quickly than those with a decline of 50–80%, which in turn recovered better than those with a decline >80%. Thus for managers, a big reduction in population size relative to historic levels should perhaps be the main factor which sets alarm bells ringing. This might be regarded as an example of 'anthropogenic rarity'; while even heavily fished species cannot usually be described as 'rare' relative to, say, white rhinos or giant pandas, it is the relative change in population size from pre-exploitation levels which is important, as much as absolute numbers.

This lack of recovery suggests a population growth rate at approximately equilibrium, as would be expected in the populations that are in the vicinity of the

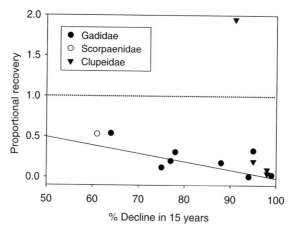

Figure 5.6. Association between the proportionately largest 15-year decline experienced by a population and the post-decline recovery, when there was some release from fishing mortality post-collapse. The solid line indicates zero recovery (population below continued to decrease, populations above showed some increase) and the dotted line full recovery; one clupeid population (69% decline followed by 1.78 recovery) omitted for clarity (Hutchings 2001).

Allee threshold. An alternative explanation is the idea of the 'predator pit', where small populations are kept at equilibrium due to heavy predation, the difference being that in this case the equilibrium is stable (Gascoigne and Lipcius 2004a, see Sections 3.2.1 and 6.5.3). (It is worth noting, however, that the biomass of large predatory fish in the modern oceans is estimated to be around 10% of pre-industrial levels; Myers and Worm 2003.) Essentially, these two different situations arise out of differences in the predator functional response, with a standard Type II response resulting in an Allee effect, and a sigmoidal Type III response (which occurs when predators neglect or are unable to locate prey which are at low density) resulting in a predator pit. In practice the overall functional response will arise from a combination of different predator species, habitats etc. This might lead to a 'compromise' between Allee effect and predator pit: a general suppression of the population growth rate at low density but with less potential for catastrophic extinction than the standard predator-driven Allee effect. These 'predator pit' type dynamics have been considered by some (perhaps have controversially) as a special case of an Allee effect which acts only at intermediate rather than low densities, and is discussed further in Chapter 6.

Fisheries for broadcast spawners
Managers need to be aware of Allee effects when dealing with a species which has a clear mechanism for a component Allee effect, such as broadcast spawning (this is not to say that all broadcast spawners suffer from demographic Allee

effects—see Section 2.2.1). Fisheries for broadcast spawners are potentially difficult to manage, as can be seen, for example, from abalone. Abalone populations in California have been serially depleted, with populations of one species after another collapsing as soon as the fishery turned their attention to them (Tegner *et al.* 1992, Parker *et al.* 1992), and there is evidence that reproductive output and population growth rate is reduced at low density (Babcock and Keesing 1999, Gascoigne and Lipcius 2004b). A sustainable abalone harvest might only be possible with highly detailed spatial management on the scale of metres, covering individual reproductive patches (Prince 2005), or via alternative techniques such as farming, which now contributes most to commercial abalone sales (Troell *et al.* 2006). This all comes too late for the white abalone *Haliotis sorensoni*, which has declined from ~1.5 million individuals to less than 2000 individuals in 30 years—a decline of ~99.9 % (Tegner *et al.* 1996, Kareiva *et al.* 2002).

Exploitation of species with Allee effects

There are, broadly, three ways in which harvesting of an exploited population can be managed. Either limits can be set on the harvesting effort ('constant effort'), or on the number (or weight) of animals harvested ('constant yield'), or else population or yield thresholds can be set beyond which exploitation is stopped (this threshold is often set at half or three-quarters of the carrying capacity; Roughgarden and Smith 1996, Stephens *et al.* 2002b). These limits can be set with various objectives in mind (minimizing extinction risk, maximizing yield, minimizing fluctuations in yield or often some combination of these) and using a variety of different estimators and modelling techniques.

The fishery for the red sea urchin in the northwestern US has been extensively studied in regard to Allee effects and best management practice, and so provides a nice case study. Models which incorporate an Allee effect via fertilization efficiency suggest, as you would expect, that management techniques which account for the spatial distribution of adults are more successful than management that just focuses on controlling effort or landings. Possible options are (i) a network of reserves, linked by larval transport (Quinn *et al.* 1993), or (ii) rotating closures (Botsford *et al.* 1993). This latter seems counter-intuitive because it opens areas to harvest where the population has been allowed to become dense to harvest, rather than protecting them permanently to ensure that the population is safe from Allee effects. Rotating closed areas will work, in theory, as long as (i) areas are never open long enough for the population to be brought close to the Allee threshold; or (ii) depleted areas are within larval transport range of a closed, dense area. It also has the benefit of avoiding reduced larval production in closed areas due to density dependent competition (Levitan 1991), however a combination of permanent and rotating closed areas might be a more risk-averse solution.

Stephens *et al.* (2002b) applied a variety of virtual exploitation techniques to a model marmot population; a species which has an Allee effect by virtue of its need to hibernate in groups for thermoregulation (see Section 2.3.1). They found the non-linear increases in extinction probability which are classically associated with Allee effects, with extinction risk being particularly acute for the constant yield and constant effort techniques when Allee effects were not taken into account. Management by threshold was 'safer' (reduced the risk of extinction) but at the expense of a highly variable harvest. They also found that limits set using limited data greatly increased extinction risk. Clearly there is no easy formula for straightforward, safe harvesting of exploited species with Allee effects; the take-home message is that conservative harvest limits and constant monitoring are vital (Stephens *et al.* 2002b).

5.2.2. **Exploitation as predation**

As well as interacting with existing Allee effect mechanisms, as discussed above, exploitation itself can act as an Allee effect mechanism, exactly like natural predation (see Sections 2.3.2 and 3.2.1). This type of Allee effect should be easier to manage in that, while hunting or fishing creates an Allee effect and an extinction threshold, the threshold vanishes if exploitation is stopped, allowing the population to recover. For as long as the exploitation continues, however, this type of Allee effect has the same dynamic consequences as 'natural' predation-driven Allee effects and can, for example, interact with a natural Allee effect mechanism in exactly the same way as other examples of double Allee effect to give rise to dormant and superadditive (or synergistic) effects (see Table 2.4, Fig. 3.5).

Exploitation leads to Allee effects because of a conflict between rational eco-logical harvesting and rational economic harvesting, because economic strat-egies are affected by a 'discount rate'. Essentially, a sum of money (a tiger skin, a box of lobsters) is worth more now than in the future, because money can be invested and will therefore generate interest in the future. Future gains are thus discounted by some percentage rate corresponding to the mean return on invest-ments. Discounting can provide an economic incentive to harvest populations at an unsustainable rate (Alvarez 1998, Lande *et al.* 1994). Indeed, exploitation to extinction is (in theory) the most profitable course of action for all populations whose growth rate is below the bank interest rate: over the long term it is more (economically) rational to exploit the entire population and invest the profits in the bank. It is economically more beneficial to cut all the trees in a forest and invest the benefits at say X% interest rate, that to sustainably harvest the forest, which grows at less than X% per year, and will thus never generate as much money. That is, of course, if all other values of this forest (from other species, for recreation, or potential yet to be discovered, to name a few) are disregarded.

If the population being exploited is large and has a high population growth rate (e.g. rabbits, anchovy), economically rational and ecologically sustainable strategies may coincide because plenty can be taken now while still leaving the population scope to replace the fraction which was harvested. However, if the population growth rate is low, the discount rate might dictate that it is economically most 'rational' to accept immediate gain over long-term sustainability, by exploiting the population to (economic) extinction, and then moving on to other populations or species in the future (Clark 1976). Hunters, poachers, fishermen, collectors etc. operating from a purely economic point of view are unlikely to reduce effort as quickly as a population declines, if indeed they reduce it at all. Mathematically, this has the same effect as either a type II functional response or a 'constant yield' type functional response from a predator (see Section 3.2.1), increasing individual mortality rate and reducing per capita population growth rate as the population declines.

An example of this type of Allee effect is shown in the exploitation of Norwegian spring spawning herring (Fig. 5.7). As the population declined due to over-fishing, catch rates remained more or less stable, because (i) fishing effort increased and (ii) herring do not become much less difficult to catch as the population declines, because of their schooling behaviour. This meant that fishing mortality increased dramatically at low population size, leading to the ecological and economic collapse of the fishery (Hilborn and Walters 1992).

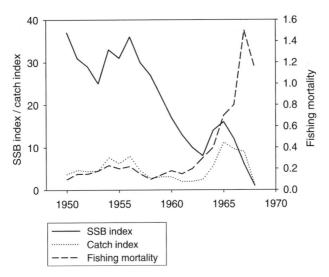

Figure 5.7. Allee effect in Norwegian spring spawning herring driven by exploitation. Population size (SSB: spawning stock biomass) declined steeply from 1957 (regression: $p < 0.0005$). However, catch rates did not change ($p = 0.15$). This meant that fishing mortality increased steeply ($p < 0.0005$). Redrawn from Hilborn and Walters (1992).

5.2.3. **Anthropogenic Allee effect**

In extreme examples of the exploitation Allee effect, rarity actually increases the value of an exploited species, creating what has been called the anthropogenic Allee effect (Courchamp *et al.* 2006). This phenomenon can occur in several types of market for rare species, including collections, luxury items (foods, skins etc.), pets, trophies, traditional medicine and even ecotourism. It arises when there are economic benefits to exploitation all the way to extinction, because as the species gets rarer its increased economic value compensates for the increased cost of exploitation as individuals become increasingly hard to find. This occurs because people regard rarity as of intrinsic value (Box 5.1). Classical economic theory, which states that at some limit the cost of exploitation of a very sparse population will overwhelm any profit to be obtained, no longer applies in this situation.

Courchamp *et al.* (2006) provide a variety of examples where rarity is linked to value (e.g. collectable Papua New Guinean butterflies, hunting trophies, exotic pets), where prices has increased with rarity (white abalone) or where spikes in demand are associated with a recognition of rarity (e.g. listing as an endangered species, see also Rivalan *et al.* 2007). In western Japan, the red morph of *Geranium thunbergii*, a flower used in traditional medicine, is common, while the white morph is rare. The frequency is opposite in eastern Japan. People in western Japan believe that the medicinal efficiency of the 'rare' white morph is better, while those in eastern Japan consider the 'rare' red morph superior in this regard. This geographic difference in people's beliefs is likely to exert strong selective pressure on flower colour and offers a good illustration of the preference for rarity and its perceived medicinal virtues (Courchamp *et al.* 2006). Some examples are illustrated in Figs 5.8, while 5.9 shows the consequence this attraction to rarity can have on the exploitation of species.

Where this type of anthropogenic Allee effect occurs, the 'rational' strategy, economically speaking, is to exploit the population as close to extinction as possible. The rarer a species, the more valuable it becomes, and the more it will be exploited, rendering it even rarer, and so on (an 'extinction vortex').

5.2.4. **Invasions and outbreaks**

Allee effects and the dynamics of invasive species
Unlike the other examples in this chapter, Allee effects are a benefit for the management of unwanted or alien species. If an invasive species suffers from Allee effects the probability of establishing a viable population is reduced, the rate of geographical spread is reduced and the area which can eventually be occupied may be smaller or patchier (Veit and Lewis 1996, useful review in Taylor and Hastings 2005, Table 3.5). Allee effects have been detected in numerous

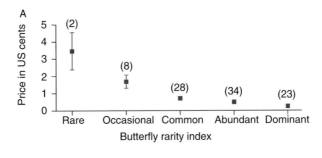

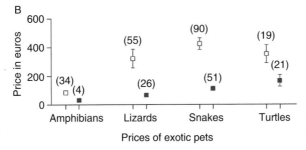

Figure 5.8. Price of collectable butterflies from Papua New Guinea as a function of rarity; standardized by dividing by male wingspan. B: Prices of exotic pet species according to their CITES status. Species listed by CITES, thus perceived as rare, are significantly more expensive than unlisted species when standardized by adult weight. From Courchamp *et al.* (2006).

invasive species, including the zebra mussel (Leung *et al.* 2004), the pine sawyer (Yoshimura *et al.* 1999), the gypsy moth (Liebhold and Bascompte 2003, Johnson *et al.* 2006), the smooth cordgrass (Davis *et al.* 2004) and the house finch (Veit and Lewis 1996); and doubtless others as well. For example, the spread of *Spartina* is slowed significantly by a pollination-driven Allee effect (discussed in Section 2.2.1), which means that plants at low density at the invasion edge can only reproduce by vegetative propagation. Models suggest that this Allee effect reduces the rate of spread of *Spartina* on the US west coast from about a 30% increase per year to a 19% increase a year (Davis *et al.* 2004, Taylor *et al.* 2004).

Several of these species have turned out to be highly damaging, so clearly Allee effects only help so far—at the beginning of an invasion when populations are still contained with low population growth rate (Veit and Lewis 1996, Cappuccino 2004). Weak Allee effects might explain the apparently common phenomenon where an introduced species is present at innocuously low density for many years before dramatic outbreaks and rapid geographical spread suddenly occur (Leung *et al.* 2004, Drake and Lodge 2006). Under these circumstances, models which fail to take account of Allee effects underestimate the risk

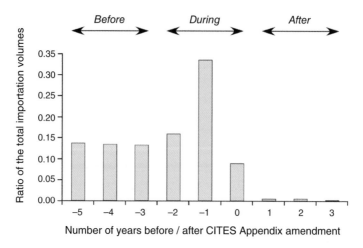

Figure 5.9. Wild collected legal importation gross volumes of 46 animal species uplisted from Appendix II to I from 1980 to 2003 according to the number of years elapsed since uplisting. For the statistical analyses, years are grouped in three time periods of three years each: '*Before*', '*During*' and '*After*' uplisting. The annual volumes were expressed as a fraction of the total volume traded over the whole period. Similar effects are found when analysing data from illegal trade, although the trade peak is then slightly later. From Rivalan *et al.* (2007).

of the invasion spreading (Leung *et al.* 2004). Models which shed further light on this process are discussed in Section 3.5.3.

Pest outbreaks

Abrupt outbreaks of pest species, whether native or alien, have been attributed to Allee effects working in reverse. The argument runs that at non-outbreak densities, an Allee effect is present, probably weak (i.e. with a decline in fitness at low density but no critical threshold) but keeping the per capita population growth rate low and maintaining the population at low density. Some event (an increase in resources, a warm summer, a decline in predators) allows the population to increase and break away from the low density at which the Allee effect operates, the population growth rate increases dramatic and a catastrophic outbreak results (Dulvy *et al.* 2004). An alternative explanation is that pests are in a 'predator pit', with Type III predation keeping them at a lower stable equilibrium (see Sections 3.2.1 and 6.5.3). In practice it is probably difficult to distinguish the two mechanisms in the field.

The highly damaging outbreaks of the coral-eating crown-of-thorns starfish on Indo-Pacific coral reefs have been hypothesized to arise due to a release from a predation-driven Allee effect or predator pit. In a comparative study of coral reef

---

Box 5.1.   **People value rare species more**

'This is one of the modern handicaps of small numbers; let a species or race become known to be rare, and museum collectors feel it their special duty to get a good supply laid in, just in case it does become extinct.' W.C. Allee, The Social Life of Animals, 1941 edition, p. 95.

Angulo, Courchamp and co-workers (unpublished data) wanted to test whether people put an inherent value on rarity, all else (size, colour, entertainment value) being equal. They designed a set of experiments which they carried out on visitors to a Paris zoo, using three different methods of assessing value: time, effort, and risk.

Time: Visitors were presented with two identical aquaria with a sign indicating a rare frog in one and a common frog in the other, but with an identical photo. They found that when the frog was present, people spent longer observing the 'rare' species, and when the frog was said to be present but was not, they spent longer looking for the 'rare' species. They also designed a website which promised a slideshow of either common or rare species, which actually did not work—people waiting for the rare species gave up later than those waiting for common species. People also persevered more with the rare species, trying more times to open the slideshow than for common species.

Effort: Visitors were asked whether they would be willing to climb up flights of stairs to look at a rare versus a common species, and if so how many storeys they would be willing to climb. 75% of people were prepared to climb as far as the third floor to see a rare species, compared to less than 50% for a common species. When put into practice, people were actually willing to spend more such effort on rare species.

Risk: A sign was put at a natural fork in the zoo's paths, indicating that one way led either to a rare or a common species, while the other led to the rest of the visit. The path to the single species was blocked by a watering hose such that visitors could not pass without getting wet. More people were willing to get wet for the rare species than for the common species. They also presented a display of two pots full of seeds, with a sign indicating that while the seeds, and the plants themselves, looked identical, one pot of seeds was from a rare species, while the other was common (in fact, they really were identical). Despite the display being almost out or reach and in full view of the public, several hundred of the seeds were stolen in a week, twice as many from the 'rare' pot.

health across a series of islands in outlying parts of Fiji, Dulvy *et al.* (2004) found that islands with higher fishing pressure had lower densities of large predatory fish, higher densities of starfish and lower coral cover. They suggest that large predatory fish usually maintain starfish at low density via a weak Allee effect. As predation rates decline with increased fishing, the starfish are increasingly able to break free from low densities where the Allee effect operates, perhaps with some trigger such as an influx of nutrients from storms or perhaps just through random demographic fluctuations.

Eradication strategies

Allee effects can simplify efforts to eradicate invasive species, as well as diseases (Garrett and Bowden 2002, Chidambaram *et al.* 2005). Generally speaking, eradication models with no Allee effects suggest that control efforts should focus on low density outlying populations which have high population growth rates and are the 'leading edge' of the invasion (e.g. Higgins *et al.* 2000). If Allee effects are present, however, low density outlying populations will not have high population growth rates, so the optimal eradication strategy might be different. With a weak Allee effect, their population growth rate would still be positive, so these populations would still have to be eradicated. With a strong Allee effect, conversely, low density populations should go extinct by themselves if cut off from a source of immigrants. Thus a strong Allee effect might alter the optimal eradication strategy qualitatively, rather than simply making eradication easier.

One example is the gypsy moth population spreading through the US, which has a strong demographic Allee effect, with a mean threshold of ~17 moths per trap (see Box 5.3, Johnson *et al.* 2006). Populations along the range edge below this density are only maintained by immigrants. The result is that the invasion proceeds in a series of pulses, with populations along the edge of the invasion front initially growing slowly due to immigration, before passing the Allee threshold and increasing to high outbreak density, whereupon emigration to new areas is likely. The current containment programme focuses on trying to eradicate low density populations in new areas, but the strong demographic Allee effect in these populations means that suppressing outbreaks along the invasion front might be key to more effective control, since this would reduce the rates of immigration into these low density new populations, keeping their population growth rate low or negative (Johnson *et al.* 2006).

As an intriguing aside, the Allee threshold in the gypsy moth is not consistent in space or time, with a lower threshold in Wisconsin than Virginia and in 2002–3 than 2003–4. The geographical variation in the threshold was, as expected, strongly correlated with geographical variation in the invasion rate (Whitmire and Tobin 2006, Tobin *et al.* 2007; Fig. 5.10). The Allee mechanism seems to be related to mate finding (Sharov *et al.* 1995, Tcheslavskaia *et al.* 2002),

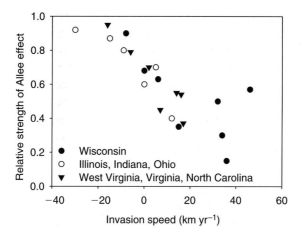

Figure 5.10. Invasion speed by the gypsy moth declines with an increase in the strength of the Allee effect; negative invasion speed indicates that the population range is contracting (Tobin *et al.* 2007).

so presumably the effectiveness of mate-finding (pheromone transmission) varies depending on the environment (or another mechanism is also present). Also, the optimal environment for gypsy moths at high densities is not necessarily the best at low density—Virginia and North Carolina had a higher Allee threshold than Wisconsin (i.e. was a less good environment for low density populations) but also had an estimated carrying capacity more than twice as high (i.e. was a better environment for high density populations) (Tobin *et al.* 2007).

Spartina, conversely, has a weak demographic Allee effect. The high density main bed produces all the propagules, and is hence responsible for long-distance transmission, but the outlying populations can still spread locally via vegetative growth. This means that both population types have to be eradicated, but the optimum eradication strategy depends on the balance between the two competing density dependent processes: seed set (positively density dependent) and vegetative spread (negatively density dependent due to competition) (Taylor and Hastings 2004).

It turns out that the optimum strategy depends on the budget that is available. If it is consistently high, the optimum strategy is to start with eradication of the high density meadow, since this reduces number of propagules and thus the risk of long distance transmission. However, if the budget is low or unpredictable, it is more efficient to concentrate initially on the low density patches, to avoid them spreading faster than eradication attempts can keep up. An intermediate budget leads to an optimum strategy which is a balance between these two processes. In this case the weak Allee effect only alters the optimum eradication strategy if enough money is available to eradicate a significantly greater area each year than is covered by vegetative spread of low density patches (Taylor and Hastings 2004; Fig. 5.11).

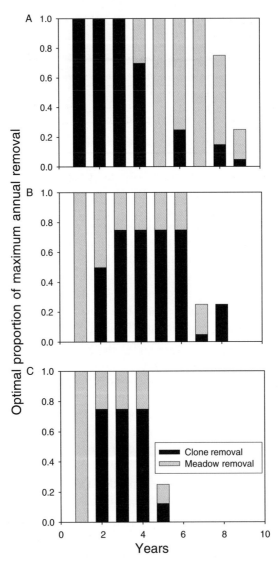

Figure 5.11. Optimal strategies for eradication of an established *Spartina* population for a low, medium or high annual budget (A–C), corresponding to the removal of 22%, 30%, or 40% of the initial area per year. A high budget favours the removal of meadow to reduce the risk of long distance seed dispersal, while a low budget favours the removal of patches to reduce the rate of short distance vegetative spread (Taylor and Hastings 2004).

Allee effects can speed up eradication, so the next step is to induce Allee effects in a population, in order to control or eradicate it. This can be done by introducing a predator or parasitoid to create a predator-driven Allee effect. We have already discussed Allee effects in biological control from the perspective of introducing small populations of the control species (Section 5.1); here we consider Allee effects in the species we are trying to control.

Theory tells us that eradication can only be achieved by a predator- (or parasitoid-) driven Allee effect if the control agent has Type II functional response; essentially if it does not switch away to alternative prey when target species density gets low (Gascoigne and Lipcius 2004a; see Section 3.2.1). Parasites can also induce an Allee effect in hosts because the number of parasites per host may be much higher when host densities are low (Fowler and Baker 1991) and this may have a disproportionate effect on the per capita population growth rate of the host population at low densities (Borowicz and Juliano 1986).

Another strategy for inducing an Allee effect in insects is to disrupt fertilization by releasing sterile males into the population, creating a mate finding Allee effect (Lewis and van den Driessche 1993, Krafsur 1998, Fagan *et al.* 2002). Models of sterile insect release suggest that it creates a density threshold in the population below which population growth rate is negative (i.e. a strong Allee effect), with the size of the threshold depending on the number or density of sterile insects maintained in the population or on the rate with which they are released (Lewis and van den Driessche 1993, Box 3.2). This technique has been around for a long time and has been successful in biological control of various pest insects (e.g. Baumhover *et al.* 1955, Iwahasi 1977, Proverbs *et al.* 1977, Graham 1978, Miller 2001). A related technique is to release sex pheromones that saturate the environment, confuse males and disrupt their response movements to females (Dennis 1989, Miller 2001). Again, models suggest this mechanism can create or increase an Allee threshold by disrupting the process of mate search (Berec *et al.* 2001).

5.3. **Detecting Allee effects**

5.3.1. **Allee effects and the Precautionary Principle**

It is perhaps useful to start this section with a general principle; the so-called Precautionary Principle set out is Principle 15 of the Rio Declaration.[3] This runs as follows: '*Where there are threats of serious or irreversible damage, lack of full scientific certainty shall not be used as a reason for postponing cost-effective measures to prevent environmental degradation.*' According to this princi-

---

[3] Declaration of the United Nations Conference on Environment and Development, Rio de Janeiro, 3–14 June 1992, full text available at http://www.unep.org/

ple, it should not matter if managers cannot confirm the presence of Allee effects in a population. Good population data is likely to be lacking on small or sparse populations, making it difficult, and even dangerous, to draw inferences about Allee effects or extinction probability. In these cases, there is a strong argument for the burden of proof to be reversed, such that an apparent absence of Allee effects should not be regarded as a proof that none is actually present (Stephens *et al.* 2007). Allee effects have such major implications for extinction probability in rare or endangered populations that if a possible mechanism exists, managers should instead assume that an Allee effect is present, or at least bear that possibility strongly in mind.

This principle provides a starting point for the management and conservation of populations with (potential) Allee effects, but it is often possible to be a little more specific; below we outline some methods for seeking out and assessing the strength of Allee effects in populations.

5.3.2. General susceptibility to Allee effects

As discussed in Chapter 4, rare species can be grouped into two categories; 'naturally rare' (occurring naturally in small or sparse populations) and 'anthropogenically rare' (reduced to small or sparse populations by human activity); this division being somewhat analogous to the distinction between the 'small population paradigm' and the 'declining population paradigm' made by Caughley (1994). Evolutionary theory would suggest that naturally rare populations will have evolved ways of avoiding Allee effects, whereas anthropogenically rare populations, which have been large and dense throughout their evolutionary history, have had no opportunity to evolve mechanisms against Allee effects.

Thus if a population which is naturally large and/or dense has been reduced to small size or low density by human activity, and if there is a mechanism by which Allee effects can operate, these populations have the potential, theoretically, to be highly susceptible to demographic Allee effects. There are, alas, many examples of this type of populations. One only has to think of the reports of travellers of centuries past, who reported scooping up cod with a bucket from the Bay of Fundy, walking over the backs of turtles in the Caribbean, eating caviar by the ladle and watching migratory herds of bison and flocks of passenger pigeons which took many hours to pass (e.g. Jackson *et al.* 2001; see Chapter 2 for a discussion of Allee effects in cod and passenger pigeon). Table 2.2 shows that the majority of examples that we have of demographic Allee effects are associated with human impacts. Thus Allee effects are at the heart of the issues which conservationists and managers have to deal with, not only because they are a potential cause of extinction, but also because they are most likely to occur in populations which have suffered from human activity.

Of course, naturally rare species can in theory suffer from Allee effects too, since mechanisms exist which affect population size and population density separately. A species with an evolutionary history of small, dense populations might not cope with low density. Adaptations for coping with Allee effects (e.g. long distance communication) may be affected by our actions (e.g. noise pollution). In addition, anthropogenically rare species are not obliged to suffer from Allee effects. A survey of fur seals and sea lion populations found good recovery in 6 out of 7 populations reduced orders of magnitude in size by overexploitation, suggesting that Allee effects are not important in this group (Gerber and Hilborn 2001), (although a weak Allee effect driven by leopard seal predation may be hampering recovery in at least one colony; Boveng *et al.* 1998). Likewise, green turtles (*Chelonia mydas*), now perceived as rare but formerly super-abundant in shallow tropical seas (Jackson *et al.* 2001), have responded rapidly in population size to a reduction in hunting and accidental killing, with a four-fold population increase in Hawai'i in 30 years and similar increases in parts of the Caribbean (Hays 2004).

5.3.3. **Detecting component Allee effects**

It is often possible to infer at least the possibility of a component Allee effect from the life history: traits such as strong sociality, broadcast spawning or a strong tendency to aggregation provide an indication that low density is not likely to be beneficial (see Table 2.1 for a more exhaustive list).

It may be possible to monitor various traits such as reproduction, adult or juvenile survival, growth, condition or predation rates in relation to population density, population size, patch size, group size, colony size etc. If, for example, a plot of proportion of chicks surviving to fledging vs. seabird colony size gave a positive slope, this would lead to some suspicion that a component Allee effects was present in these populations (lower chick survival in small colonies). It would not provide conclusive proof, since both colony size and chick survival might be linked to a third factor (e.g. better foraging grounds close by), but the management of endangered populations does not (or should not) operate by requiring scientifically rigorous and conclusive proof of threats to their survival.

The best way to test whether a population has a component Allee effect is to conduct an experiment where replicate populations of different sizes or densities are created (e.g. Levitan *et al.* 1992, Groom 1998) or where individuals are translocated between natural populations of different size or density (Gascoigne and Lipcius 2004c). This is, however, frequently impossible in practice, since interfering with populations of endangered and protected species is usually neither permitted nor recommended. Sometimes, however, natural 'experiments' exist, as for example when there are several populations living under different conditions (e.g. Angulo *et al.* 2007).

5.3.4. **Detecting demographic Allee effects**

Demographic Allee effects are, of course, the major concern when dealing with endangered species, since it is demographic rather than component Allee effects which lead to low or negative population growth rate in small or sparse populations. Unfortunately, detecting demographic Allee effects requires data such as estimates of fitness or population growth rate over a range of densities, which is difficult to collect. However, there are a variety of techniques which may help in assessing the potential of a population for demographic Allee effects, which we outline below.

Population model fitting
The most usual way that demographic Allee effects have been detected, or at least inferred, in the field, is by incorporating a component Allee effect (verified by field observation or experiment) into a density-dependent population model (for techniques and examples see Section 3.2). Alternatively, models with and without an Allee effect can be fitted to field data to see which gives the most parsimonious description of observed dynamics (see Section 3.3). Such models can be structured as seems most appropriate to the population: by age or stage, vulnerability to predation or exploitation (e.g. Gascoigne *et al.* 2004b,c) or even by density (Taylor *et al.* 2004). The predications of models are, of course, only as good as their estimates of parameter values (rates of growth, survival and reproduction). A particular problem is often to estimate density-dependent relationships in various components of fitness. For example, a component Allee effect in mating probability may or may not be mitigated by high offspring survival at low density due to release from competition—depending on the relative strength of these density-dependent relationships (e.g. see Angulo *et al.* 2007). Thus, in terms of quantitative predictions about a specific population, this modelling approach is only useful if significant amounts of population data are available.

Observations of fitness
Another common way in which demographic Allee effects have been inferred, and the Allee threshold estimated, is from field observations which allow an assessment of individual fitness. An obvious example is plants that are potentially pollen-limited; reproductive output (pollination, seed set) can be directly monitored to see whether there is an Allee threshold below which reproduction is impaired or prevented. Groom (1998) provides a nice example in a study of *Clarkia concinna* (an annual herb), where population size and isolation interact to determine a threshold below which there is reproductive failure and a higher probability of extinction.

In animal populations, behavioural observations can sometimes allow us to draw inferences about fitness. For example, Courchamp *et al.* (2002) combined behavioural observations in the field with modelling to try and assess the minimum pack size threshold for African wild dogs (see Section 2.4). Observations suggested that packs of five adults or fewer were less likely to leave an adult as a pup guard (babysitter) during hunts than larger packs, and as a consequence suffered high pup mortality. The modelling suggested that if these smaller packs left a pup guard, they would have to increase their hunting rate significantly or face nutritional deficit. Thus both approaches suggest that five adults per pack is the Allee threshold. Behavioural observations of the desert bighorn sheep (*Ovis canadensis*) likewise suggested that in groups of five individuals or less, individual vigilance increased (imposing a fitness cost) but total group vigilance, and thus protection from predation, decreased (Mooring *et al.* 2004); it is likely that an individual's overall fitness declined in groups smaller than this threshold size.

Allee effects might also be inferred from knowledge of dispersal behaviour, particularly where dispersal is risky. If dispersal rates away from small populations are high relative to dispersal rates from large populations, this suggests that individuals may have low fitness in small populations and will take a risk (disperse) to improve it. This has been shown to occur in a colonially breeding bird, the lesser kestrel; Serrano *et al.* 2005) and the Glanville fritillary butterfly (Kuussaari *et al.* 1998). However, high dispersal rates are not always a signature of low fitness; see Section 6.2.1.

Population time series

It is possible to look for Allee effects if a time series of population size or density is available, by looking for signals of positive density dependence in the *per capita* population growth rate at low density. *Per capita* population growth rate is calculated from a time series of population size or density as the ratio between the population size N or density D from one year to the next. Calling the year of interest year t, the mean *per capita* population growth rate during that year is calculated as N_{t+1}/N_t or D_{t+1}/D_t (Liebhold and Bascompte 2003). Being a ratio, and thus dimensionless, either population density or population size can be used. Other measures of population size might also be appropriate, including number per trap, per transect, per person, day of survey etc., as long as they show a linear relationship with population size or density.

Analyses of population time series, including low densities, for mammals (Fowler and Baker 1991), birds (Saether *et al.* 1996) and fish (Myers *et al.* 1995) have, however, found very little evidence of demographic Allee effects. This may be because Allee effects are generally rare in these groups, but it may also be because population data is inevitably very noisy, and there are sundry problems and biases associated with time series analysis (see Box 5.2). Nonetheless, these analyses can be very revealing (e.g. Wittmer *et al.* 2005, Angulo *et al.* 2007).

Box 5.2. **Problems with population time series analysis**

Time series analysis poses a variety of statistical problems. One is 'autocorrelation'—the population size at time t is not independent of the population size at time $t-1$ (the assumption of independent data points is implicit in most statistical analyses). The extent to which independence can be assumed depends on the size of the time step (usually a year) in relation to the life history of the population in question. For species with long life spans and generation times, independence is often a poor assumption, but for species with short generation times and non-overlapping generations it might be more valid.

There is also an inherent bias towards detecting negative density dependence in the analysis of time series. This arises because even when population growth rate fluctuates randomly, a high value is more likely to be followed by a lower value, while a low value is more likely to be followed by a higher value, statistically speaking (Freckleton *et al.* 2006). This effect is exacerbated by measurement error, since an underestimate of the population in year t followed by a better estimate in year $t+1$ leads to an overestimate of the population growth rate during year t correlated with the incorrectly small population estimate in year t (Freckleton *et al.* 2006). Spurious negative density dependence can also arise via changes in the population age structure with density. If, for example, denser populations contain a higher proportion of older individuals with high mortality, denser populations will show higher mean mortality rates unless the data are disaggregated according to age class (Festa-Bianchet *et al.* 2003).

Another problem, particularly in marine systems, is the definition of a 'population'. The data collected for fisheries management purposes is frequently aggregated in geographical areas which have no relationship to the distribution of fish populations, but instead correspond to political boundaries (Fig. 5.12). If these data actually come from several distinct populations, Allee effects in individual populations will not be apparent in the aggregated data until the Allee threshold is reached across the whole metapopulation (Frank and Brickman 2000, Freckleton *et al.* 2006).

Last but by no means least is the problem of noise, via demographic and environmental stochasticity and sampling error, which results in an analysis with low statistical power. This is particularly a problem for detecting Allee effects, which by definition occur only over a small range of population densities or sizes (see Myers *et al.* 1995, Shelton and Healey 1999).

Box 5.2. *(Continued)*

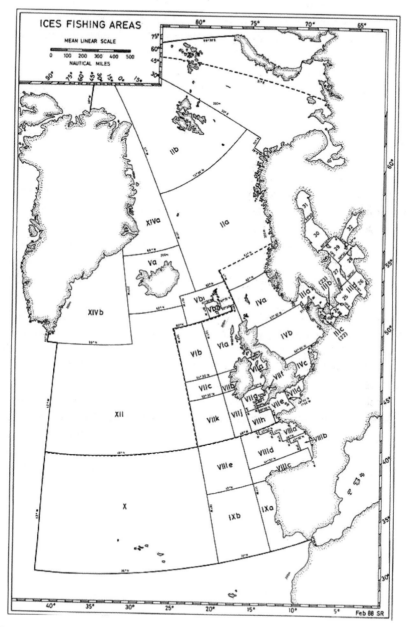

Figure 5.12. Geographic units used for fisheries management in larger Northeast Atlantic area. Fisheries managers are well aware that these areas do not generally correspond to distinct populations, but defining the true geographical extent of fish (meta) populations is not a simple task.

In a few cases, such time series analyses may be possible using a whole series of populations rather than just a single population. This approach is clearly much more robust and avoids (or at least mitigates) some of the difficulties associated with time series analysis set out in Box 5.2. In this case, the sign of the change in population size from year to year (increase vs. decrease) can be used to assess whether the population is above or below any threshold, with the threshold being defined as the population size or density at which a population has a 50% chance of persistence. This technique is probably easiest to explain using an example, and one is presented in Box 5.3.

Spatial distributions
We might predict that populations with demographic Allee effects will be patchily distributed in space, since low density populations would have a high probability of extinction. This simple prediction is complicated by dispersal—a small population with a negative population growth rate may persist because continuously supplemented with immigrants (see Section 3.5.1). Nonetheless, a 'gap' in the frequency distribution at small or sparse population size (i.e. a landscape where small or sparse populations do not exist) might be an indicator of demographic Allee effects. A highly dispersed distribution, on the other hand, might

---

Box 5.3. **Calculating the Allee threshold in invading gypsy moth populations.**

An extensive network of about 150,000 pheromone baited traps has been used annually since the mid-1990s to monitor the invasion front of gypsy moths in the northeastern USA. Tobin *et al.* (2007) used a subset of this trap data to interpolate population size in a network of 5x5 km grid cells covering the states of West Virginia, Virginia and North Carolina for each year from 1996 to 2004. They then matched the population in each grid cell at time t with the population at time $t-1$ (recall that the ratio of these is a measure of the per capita population growth rate). This gave them more than 20,000 individual pairs of moth counts in successive years.

For each integer value of population size in time $t-1$ they calculated the proportion P of populations starting at this size which increased. At the Allee threshold, $P = 0.5$, because a population at the threshold size is equally likely to increase or decrease the following year. Smaller populations are more likely to decrease because of the Allee effect, while larger populations are more likely to increase (Fig. 5.13). (There is also a higher density threshold at which $P = 0.5$; this corresponds to the carrying capacity.)

Box 5.3. *(Continued)*

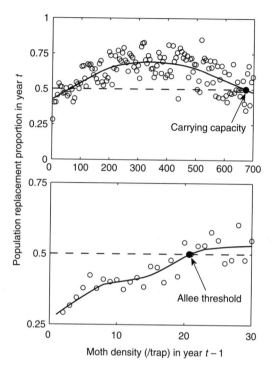

Figure 5.13. The proportional size (*P*) of the population in year *t* (y-axis) relative to the population size in year *t*–1 (x-axis). The Allee threshold at $P = 0.5$ works out at ~20.7 moths per trap, the carrying capacity at ~673 moths per trap. The bottom panel is a close-up of the low-density populations of the top one. Redrawn from Tobin *et al.* 2007

indicate a population which is adapted to life at low density. Conversely, (ironically) it might indicate a population at risk from Allee effects! As we have argued above, this will depend on the evolutionary history of the population in terms of size and density (naturally vs. anthropogenically rare), as well as whether a mechanism for Allee effects exists. In other words, such analyses are open to interpretation, sometimes in two diametrically opposing ways.

A comparative index of dispersion can be calculated relatively easily, for example as the ratio of variance to mean in population density (Pierce *et al.* 2006). A similar, if more complex, technique has been suggested for detecting Allee effects in populations which cover two different habitat types. Ideal free distribution theory suggests that individuals move between habitats to maximize their own fitness. If this applies (i.e. if the species is mobile), and an Allee threshold is present, the result is a threshold in habitat occupancy in the less preferred habitat, with only the preferred habitat occupied at low density (Greene and Stamps 2001, Morris 2002). Even if the patches are of equal quality, only one will be occupied at low density, although in this case it is impossible to predict which. Applying this technique to populations of red-backed voles (*Clethrionomys gapperi*) and deer mice (*Peromyscus maniculatus*) in two vegetation types in the Canadian Rockies, Morris (2002) suggested that Allee effects might be present in both species, with a threshold of ~13 voles and ~5 mice per hectare in the lower-fitness habitat.

Experiment

It is simpler to demonstrate an Allee effect in populations where an experiment can be carried out. The bush cricket was introduced into habitat patches in groups of 2, 4, 8, 16, and 32 individuals (1:1 sex ratio). Population size and probability of persistence depended significantly on initial propagule size, and the threshold, measured as the propagule size at which there was a 50% probability of persistence, was ~16 individuals (Berggren 2001, see Fig. 5.14). The mechanism for this possible Allee effect remains unclear, since a previous study suggested that the mate-finding ability of this species is always high, even at very low density (Kindvall *et al.* 1998).

In another experiment, Fauvergue *et al.* (2007) assessed the existence of Allee effects in a parasitoid wasp *Neodryinus typhlocybae* (the main control agent in

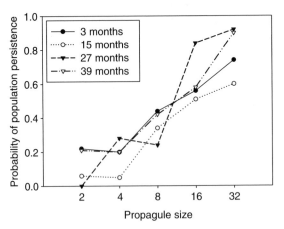

Figure 5.14. Probability of persistence of experimentally introduced populations of bush crickets in relation to initial propagule size. Propagule size at a persistance probability of 0.5 (~16) is an estimate of the threshold (Berggren 2001). Note that the individuals in the population were not always detected at each time period, resulting in some populations 'reappearing' in the dataset after going extinct.

France for an invasive crop pest) by releasing 45 populations made of 1, 10, or 100 individuals. Initial population size did not affect population establishment probability, growth rate or extinction probability: no Allee effect was present. As a result, the optimal release strategy in this species is as many releases of as few individuals as possible.

Several experiments have evaluated fertilization efficiency as a function of density or nearest neighbour distance in broadcast spawning marine invertebrates (see Fig. 2.2). A combination of fertilization experiments in the lab, measurements of fertilization efficiency in the field and modelling, suggests that in the greenlip abalone (*Haliotis laevigata*) a critical threshold for recruitment failure in the field of ~0.3 animals m^{-2} (mean nearest neighbour distance 1–2 m) corresponded to a fertilization efficiency of ~50% (Babcock and Keesing 1999); it would be interesting to investigate whether this is a general rule.

5.4. **The short version**

We have tried in this chapter to provide a thorough and exhaustive discussion of the role that Allee effects play in different aspects of conservation and management. Perhaps, however, the discussion is a little too exhaustive for busy people at the sharp end of conservation—those taking day-to-day management decisions about the management of endangered or exploited populations. For these people we provide below a brief primer summing up the conclusions of this chapter.

Should managers and conservationists be concerned about Allee effects, even in populations which are not apparently in immediate danger? Yes they should, because Allee effects (i) can create critical thresholds in size or density below which a population will crash to extinction; or (ii) even if there is no threshold, they increase extinction probability (due to stochasticity) and reduce per capita population growth rate in low density or small populations.

It is hard to confirm or refute definitively the hypothesis that a species has Allee effects; in many cases it is probably not worth trying. But there are some conditions under which an Allee effect is more likely, as well as some indications that it might be present:

- The species is a broadcast spawner, mast seeder or otherwise dependent on 'group action'

- The species naturally occurs in dense aggregations, herds or flocks

- Mate-finding is likely to be a problem

- The species is a cooperative breeder or relies on complex social behaviours

- Predation is an important source of mortality; the main predators are generalists (perhaps introduced); anti-predator strategies involve aggregating, herd behaviour or vigilance

- The species naturally existed in stable, large or dense populations which have now been greatly reduced in size and/or density

- There are indications that small and/or sparse populations do not increase in size or density when protected from whatever reduced them to small size (exploitation, introduced predators etc.)

- Populations have been recorded as going extinct abruptly or unexpectedly, or crashing rapidly to small numbers

- Repeated attempts at reintroduction have failed without obvious external cause

What should you do if you are concerned about Allee effects in populations you are trying to manage? The obvious answer is to maintain the population above the extinction threshold. This is easier to say than do especially since the extinction threshold is difficult to estimate and will vary according to circumstance—habitat quality, mortality rates from predator or exploitation etc. Experience may show that populations below a certain size are not usually viable, providing a minimum estimate. Knowledge of a healthy expanding population may give an upper bound, although not one which can necessarily be assumed in different habitats or situations. More generally, a consideration of Allee effects should be incorporated into any target which is set for the population size, whether that be set using PVA or some other criterion; for example, Allee effects may act to reduce the range of a population, and thus their ecological function in certain areas.

Allee effects interact synergistically with mortality from exploitation, which not only reduces the population size but can also create or increase an extinction threshold. Exploitation of populations with possible Allee effects need to be considered and managed very carefully. Exploitation can also act as a mechanism to create an Allee effect, particularly if the economic value of a species is found to be increasing with rarity (the anthropogenic Allee effect). By the time this situation is reached, however, conservationists will probably be all too aware of the danger.

Allee effects alter optimal strategies for (re)introductions. Without Allee effects, it is preferable to spread introductions out in time and space, to reduce risk. With Allee effects, however, it is important to maximize the initial size of the introduced population as much as possible. Conversely, for eradication of populations with Allee effects, it may prove more efficient to focus on high density areas rather than outlying low density populations, as is usually the case.

Generally, if Allee effects are suspected in a given species, it may be most efficient (if financial resources are scarce) to direct conservation efforts at maintaining large or dense populations, rather than at trying to preserve and increase small, sparse populations which may no longer be viable (translocation or reinforcement may be useful tools). At the habitat level, this probably means protecting large, central fragments at the expense of small, isolated or peripheral fragments.

6. *Conclusions and perspectives*

6.1. **What you have just read, and what awaits you now**

Unless you cheated a bit, you have now gone through some 200 pages of ecological marvels and in the process you have become an expert on Allee effects. In Chapter 1 of this book, you discovered how the pacifist principles of a great scientist generated the idea of Allee effects in the first place, and how the science of population dynamics has taken half a century to mature from a competition to a cooperation-based mindset. In Chapter 2, in addition to all the mechanisms generating Allee effects, you also learned why a sessile life history is uncommon for land animals and why the emergence cycle of cicadas often has the period of a prime number. With Chapter 3, you acquired the ability to impress friends and colleagues by mastering imposing equations as well as cool terms such as phenomenological models of demographic Allee effects, sigmoidal functional responses, double dormancy and superadditivity. Thanks to Chapter 4, you now understand why Allee effects might be more common in naturally large and dense populations than in naturally rare ones, and why Allee effects may have led to adaptations as diverse as sperm storage, pheromone production and life-long monogamy. In Chapter 5, you saw how Allee effects can be both an ally and a foe in turn, thereby helping you save white abalones and black lemurs from extinction and fight invasions of gypsy moth and smooth cordgrass. You could very well end this reading right here, because the large majority of important things has been said. It may be worth continuing, however, since in this final chapter, we will exploit this newly acquired insight into ecology and conservation to discuss some thought-provoking, exciting and (perhaps) controversial Allee effect related matters.

We will first delve more deeply into the problems of studying Allee effects, touched on slightly in previous chapters. In the same vein, we will discuss issues where Allee effects have got controversial, including the relationships between demographic stochasticity and Allee effects. We will also consider briefly some matters that concern the core idea of the book less closely, but that are none-theless connected to Allee effects, and that are intellectually stimulating: these include Allee effects and ecosystem shifts and Allee-type effects in some other scientific disciplines. Finally, we will attempt the perilous but fascinating task of predicting the future, proposing some avenues for new research.

6.2. **Problems with demonstrating an Allee effect**

As we saw in Chapters 1 and 2, and despite our eagerness to promote Allee effect recognition, it is important to keep in mind that the existence of a mechanism known to create an Allee effect does not systematically imply a component Allee effect, and that the existence of a component Allee effect does not systemati-cally translate into a demographic one. In addition, even if a demographic Allee effect exists, it may be difficult to demonstrate it because small populations are vulnerable to extinction for various reasons, including different types of Allee effect, but also vulnerability to catastrophes, environmental stochasticity and demographic stochasticity, and of course anthropogenic factors (Lande 1998a, Oostermeijer *et al.* 2003).

6.2.1. **Allee effects and confounding variables**

Demonstrations of a positive relationship between individual fitness and popu-lation size or density in natural populations may not provide completely con-vincing evidence for Allee effects. A fundamental problem in field ecology is that while observational evidence can show a correlation between two variables (population size or density and individual fitness in this case) only experimental manipulation can allow to infer causality—that decreased fitness is *caused* by decreased population size or density. However, observational data is frequently the only data available, particularly when dealing with rare, protected species, or populations such as vertebrates which are a challenge to experiment on, to say the least.

Confounding variables are thus a major issue with observational studies of Allee effects. Large and small populations may differ in a variety of ways other than size—habitat type or quality, microclimate, resources, disturbance and so forth. Recall we mentioned in Chapter 1 that a positive correlation between fit-ness and population size may lack causality and be instead only generated by habitat quality. This is not an Allee effect, even if it looks like one. It is important to bear in mind that not all field studies which purport to show Allee effects take account of confounding variables, so a little scepticism is vital when exploring the scientific literature. (The flip side of this is that there are some studies which demonstrate an Allee effect without even mentioning it.)

Dispersal is another issue that can confound apparent observations of Allee effects, despite the fact that variable dispersal rates and dispersal mortality can be a cause of Allee effects, as in the lesser kestrel (Serrano *et al.* 2005; Fig. 2.11). In general, however, high dispersal rates are open to different interpretations. A study of rabbits in Australia, for example, found that pre-reproductive individuals were more likely to disperse long distances away from their home warren when the warren was reduced in density, with a threshold for staying put at 60–75 adults/km^2 for females and ~90 adults/km^2 for males (Richardson *et al.* 2002). This on the face suggests an Allee effect, since dispersal is usually risky—if individuals risk dispersing at low density, this suggests that the fitness they obtain by staying put is low. In fact, however, warrens which were reduced in density subsequently received high numbers of immigrants, more than compensating for losses due to emigration. It turns out that high dispersal rates at low density are related to ben-efits of gaining mating opportunities in other warrens, with juvenile rabbits mak-ing exploratory journeys to evaluate risks and benefits before choosing to disperse (Richardson *et al.* 2002). In other words, while in the kestrel, dispersal away from small colonies seems to be associated only with high costs of staying put in small colonies (an Allee effect), in rabbits, dispersal away from low-density warrens is associated with reproductive benefits to be gained in new low-density warrens (higher dominance, lower relatedness etc.). In other words, the decision to disperse at low density may be based on costs associated with low density in the present location, or may be based on knowledge about the benefits at a new location.

A related issue concerns studies based on patch size. Field studies on species living in fragmented habitats quite often focus on habitat patch size rather than directly on population size or density. Clearly, the two are not necessarily the same, since a small (sparse) population could inhabit a large habitat patch, while a large (dense) population could in theory also live in a small patch. In practice, however, the two are usually positively correlated (see Box 2.2). However, the distinction between patch size and population size or density in relation to Allee effects can be a bit more subtle. It is nicely illustrated by the case of the northern spotted owl, a species reliant on the fragmented remains of old growth forest in the north-western USA.

At first sight, the owl looks again like an excellent example of a species with an Allee effect, since mate-finding is significantly impaired in small habitat frag-ments. In large areas of habitat with clusters of suitable territory sites, males can easily find suitable empty territories, and females can then find territories occupied by single males. Conversely, in small, isolated patches of habitat, suit-able territory sites are far apart and separated by unsuitable matrix habitat. Under these circumstances owls experience high dispersal mortality and have a low probability of finding a suitable territory. A combination of field work and population modelling suggests that there is a threshold habitat patch size corresponding to about 15–20 suitable territories, below which the population in that patch will go extinct unless it is continually supplied with dispersing owls from other patches (Lamberson *et al.* 1992, 1994, McKelvey *et al.* 1993).

This certainly sounds like an Allee effect. However, although there is a correlation between population size (via habitat fragment size), mean individual fitness (via mate finding) and population extinction risk, an Allee effect is only present if the decline in individual fitness or per capita population growth rate is caused by the decline in population size itself, rather than both being caused by habitat fragmentation. The test is to ask whether the Allee-type effect would still occur if a population was reduced in size or density in the absence of habitat fragmentation.

In this case, the answer is: probably not. Given a small population in a continuous forest, males would have no difficulty finding new territories round the edge of the existing population range, and females no difficulty finding males—so no Allee effect. Likewise, in a low-density population males would again have no trouble finding territories. Females might have more difficulties finding males, but since they would not be forced into unsuitable matrix habitat, dispersal mortality would be lower. Thus the Allee-type effect here is likely to be a consequence of habitat patch size, not population size or density, and is thus not really an Allee effect.

In Box 2.2 we argue that for pollen limitation in plants, habitat patch size is a reasonable proxy for population size or density, and can lead to a component Allee effect. Are we contradicting ourselves here? Don't worry, we're not. In vector-pollinated plants, there are at least two interacting mechanisms for Allee effects: (i) reduction in pollinator visits and (ii) reduction in supply or quality of conspecific pollen. A large or dense population in a small habitat patch would suffer from (i), a small or sparse population in a large habitat patch would suffer from (ii) and a small or sparse population in a small habitat patch would suffer from both; i.e. habitat patch size and population size interact with each other. This is not the case for northern spotted owl, where the mate-finding problems are driven directly by habitat fragmentation, not population size *per se*.

This distinction might sound pedantic, but in fact it might make a difference to the management strategy. Owl populations around or below the threshold size can in theory be saved by increasing habitat patch size and thus the number of suitable territories (perhaps by designating larger reserves or managing the surrounding matrix so it is less hazardous for owl dispersal). This would not be the case if the owl had a true Allee effect—such populations would be likely to go extinct regardless of increases in habitat availability, and could only be managed by augmenting both the habitat and population size.

6.2.2. **Allee effects and demographic stochasticity**

An issue which has exercised ecologists a fair amount over the last few years concerns the relationship between Allee effects and demographic stochasticity, and specifically, whether demographic stochasticity can be included in the list of Allee effect mechanisms or not (Lande 1993, 1998b, Stephens *et al.* 1999, Engen *et al.* 2003, Bessa-Gomes *et al.* 2004). Population ecologists currently distinguish two types of demographic stochasticity: (i) random fluctuations resulting from individual birth and death events, and (ii) random fluctuations in the adult sex ratio in populations with two sexes.

Recall that according to the definition of an Allee effect in Chapter 1, demographic stochasticity (of any type) can be considered an Allee effect mechanism only provided that it reduces mean *individual* fitness at low population size or density. This clearly does not apply to random fluctuations resulting from individual birth and death events, since the chance events drive individuals out of the population or allow them to produce a given number of offspring with a density-independent probability. It is only simultaneous bad luck in a number of conspecifics that can drive the entire population to extinction. This is obviously more likely when there are fewer individuals: as a result, the probability that the population size collapses to zero increases as the population declines. However, because Allee effects concern individual fitness, demographic stochasticity in births and deaths cannot be classified as an Allee effect mechanism, even though it increases extinction risk of the population as its size decreases (Stephens *et al.* 1999, Bessa-Gomes *et al.* 2004).

On the other hand, sex ratio fluctuations arguably do reduce mean individual fitness as the population declines. They arise both as a consequence of chance in determining sex of the offspring and because of demographic stochasticity in male and female deaths which, even if the male and female death rates are equal, causes male-to-female ratio to vary unpredictably. To explain how this relates to individual fitness and population size, we recapitulate here the argument given by Stephens *et al.* (1999). Consider a species with a 1:1 sex ratio at birth. When a population of this species is composed of only two individuals, they can be both male, MM; one male and one female, FM; or both female, FF. These possibilities arise with probabilities p(MM) = 0.25, p(MF) = p(FM) = 0.25, and p(FF) = 0.25. The probability that a reproductive pair form is thus p(MF) + p(FM) = 0.5 and hence the probability than an individual is in a position to reproduce is also 0.5. With three individuals, however, the probability that an individual cannot mate p(MMM) + p(FFF) is only 1/9 + 1/9 and thus the probability that they can mate is 7/9; with four individuals it is 7/8 etc. In other words, as the population size increases, the adult sex ratio is less and less likely to deviate from the mean sex ratio 1:1 and nearly every individual within the population will be able to find a mate. Because demographic stochasticity due to random fluctuations in the adult sex ratio reduces a component of individual fitness—the mating rate—as population declines, it can be reasonably considered an Allee effect mechanism (Stephens *et al.* 1999, Bessa-Gomes *et al.* 2004).

This kind of Allee effect is more likely to occur in monogamous than polygynous populations (Engen *et al.* 2003, Bessa-Gomes *et al.* 2004), and has been modelled in marmots (Stephens *et al.* 2002a) and measured in small plant populations (Soldaat *et al.* 1997), although in neither case were individual fitness or per capita population growth rate measured directly. Another case, widely advertised at the time, is the kakapo (*Strigops habroptilus*), a lek-breeding giant, flightless parrot native to New Zealand. In 2001, the world population of this flightless bird consisted of 54 individuals, of which only 21 were female (with few of these being fertile), distributed across several islands. An intensive breeding program has resulted in the birth of several chicks, but only six female fledglings have been produced since 1982 (Elliott *et al.* 2001, Sutherland, 2002). Typically, the fewer kakapo breed, the more likely can random fluctuations in the adult sex ratio lead to a dramatically male-biased population, with little possibility of recovery.

6.2.3. **Allee effects and sex ratio biases**

Another mechanism by which sex ratios can be altered is selective exploitation, when hunters or fishermen preferentially select either males or females; generally males.

Ecologists usually regard males as less important to population dynamics than females, since one male can often inseminate a large number of females (but see Rankin and Kokko 2007). Nonetheless, a striking example of reproductive failure under selective exploitation for males can be found in the saiga antelope (*Saiga tatarica tatarica*) from the Central Asian steppe, whose males are selectively poached for their horns, for Chinese medicine. The recent political upheaval has lead to a dramatic increase in poaching, and the population size has declined at an average annual rate of 46%, with adult males declining even more rapidly. Saiga antelope has a polygynous mating system, so the population can support a significant female bias without any decline in female reproductive output. In 2000, however, the population reached a critical threshold in the number of males, with female reproductive output falling dramatically as the proportion of males fell below one male for every 36 females (Milner-Gulland *et al.* 2003).

A similar situation can arise in fish which have complex reproductive systems. Sex change is common, and can occur as both protogyny (from female to male) or less commonly protandry (from male to female) (Sadovy 2001). In these species, fisheries which target the largest individuals take almost exclusively one single sex. Sex change is socially mediated, so to some extent fish populations respond to changes in the sex ratio via increases in the rate of sex change (Warner 1984, Shapiro, 1989), but this may not happen quickly enough

Figure 6.1 The dramatic shifts in adult sex ratio imposed on the saiga antelope *Saiga tatarica tatarica* have led to a population threshold below which there is reproductive collapse. Photo: Anna A. Lushchekina.

to counteract heavy fishing pressure. In gag grouper (*Mycteroperca microlepis*), adult sex ratios changed from 17% to 2% male due to selective fishing for males in spawning aggregations (Koenig *et al.* 1996). This process has led to population collapse, for example in the red porgy, *Pagrus pagrus*, where the reproductive output of the population fell below 1% of the estimated unfished level, due to decreased population size and skewed sex ratio (Huntsman *et al.* 1999).

The dramatic shifts in adult sex ratio seen in saiga antelope and reef fish, and their consequences for reproductive output, conform with our idea of Allee effects in the sense that the population reaches a threshold below which there is reproductive collapse. As with the spotted owl example, however, it is arguable whether there is a direct causal link between sex ratio shifts, low density and reproductive collapse. In species where mate-finding is roughly a random process, clearly skewed sex ratio would interact with low density to create a stronger Allee effect than low density alone. However, saiga antelope and most reef fish have behaviours (harems, spawning aggregations) which avoid low density and hence direct mate-finding problems even in small populations. Thus the problem is purely one of sex ratio, which could in theory drive reproductive collapse at any population size. Reproductive collapse arises not directly from an Allee effect, but from a non-linear threshold switch between two alternative stable states: unharvested, where the effective population size is determined by the number of females, and harvested, where below the critical sex ratio threshold the effective population size is determined by the number of males. We discuss this idea of alternative stable states in more detail below.

6.3. **Allee effects and ecosystem shifts**

Allee effects are just one manifestation of a positive feedback loop in ecological systems. Positive feedback loops arise out of positive or facilitative interactions between individuals, whether within the same species (an Allee effect) or within ecological communities, food webs or ecosystems. These types of interactions, as we have seen with Allee effects, are inherently destabilizing for dynamics. Other examples of ecological systems driven by positive feedbacks include pest outbreaks (discussed in Section 5.2.4), patterning and transitions in space such as treelines, tiger bush and patchy mussel beds (Box 2.4) and alternative stable states in ecosystems. The latter are sometimes called ecosystem shifts or regime shifts (Carpenter 2003). All these processes have a common driving mechanism— positive interactions within or between populations which create abrupt thresholds between one state and another, either in space or in time (e.g. persistence vs. extinction, outbreak vs. non-outbreak or vegetation vs. bare substratum). We have already stressed how ecologists have focused on negative feedback mechanisms which regulate and stabilize dynamics (competition in the case of populations), and how the role of positive feedbacks in structuring communities as well as populations has been to some extent overlooked, as well as being fragmented into these specific cases without an overarching theoretical framework.

This book is not the place to develop such a framework, but we would like to consider briefly the connection between Allee effects (in populations) and threshold transitions between stable states in ecosystems. There are several well-known examples of such transitions in ecosystems, analogous to the 'switch' between extant and extinct in single species models with Allee effects. They seem to arise mainly in systems where some form of vegetation is controlled by a herbivore, which in turn can be controlled by a predator or suite of predators ('top-down control'). A perturbation at one level in the food web can have knock-on effects at other levels, triggering the failure of top-down control at some point in the chain, a 'trophic cascade' and a switch to an alternative ecosystem state not characterized by top-down control. This alternative state is 'stable' if the removal of the factor causing the perturbation is not sufficient to switch the system back to its old state. For instance, an increase in nitrate concentration in a pond over a threshold of X might trigger a switch to an alternative state (dominated by different species of plants and animals) but a reduction in nitrate back below the threshold of X will not be sufficient to trigger a switch back to the original state. This asymmetry between the route from State A to State B and the route back from State B to State A is called 'hysteresis'. Extinction is perhaps the ultimate hysteresis, since the transition between extant and extinct is only possible in one direction.

A very simple example which bridges the gap between populations (Allee effects) and communities (ecosystem switches) can be found in beds of mussels (*Mytilus edulis*) on soft sediment, where the analogy with the Allee effects is obvious. Mussels do not usually occur on soft sediment because of a lack of suitable attachment sites. When put there (usually by mussel farmers) they can attach to each other, and grow well. Once established, these mussel beds can persist indefinitely, creating a new type of ecosystem with a different sediment type and a different suite of associated species (Beadman *et al.* 2004). If lost, however, the mussels would not be able to re-establish without human intervention (i.e. there is a hysteresis). This highly simplified 'ecosystem' thus has two alternative states: mussel bed vs. bare substrate, each stable once established, with the switch mediated by an Allee effect in the mussels which have low survival at low density. In fact, viewed in this light, any Allee effect can be thought of as leading to an ecosystem switch when viewed from a community-wide perspective (regardless of whether it causes secondary extinctions or enables other species to invade).

More complex examples arise in aquatic systems such as lakes and coral reefs. Freshwater lakes, for example, can buffer nutrient enrichment over a wide range if they support extensive stands of submerged macrophytes, which compete with phytoplankton for nutrients, stabilize sediments (preventing nutrient release) and provide a refuge for zooplankton grazers from fish predators. Large zooplankton populations can then graze down phytoplankton, maintaining clear water even in the face of high nutrient concentrations. However, if nutrient enrichment leads to excessive phytoplankton growth, this can in turn lead to light limitation and loss of macrophytes. The lake ecosystem can then switch to an alternative, phytoplankton-dominated state. In this case, light limitation impedes the macrophytes from re-establishing, while heavy predation by fish in the absence of a spatial refuge prevents zooplankton grazing from regaining control of phytoplankton populations. Thus even when nutrient inputs are reduced to low levels, the ecosystem cannot automatically revert back to its original state unless planktivorous fish populations are controlled (Scheffer 1990, Scheffer *et al.* 2001, Gulati and van Donk 2002, Irfanullah and Moss 2005).

In this situation, phytoplankton can be regarded as a 'pest' species which outbreaks in an analogous way to the crown-of-thorns starfish (Section 5.2.4), when released from an Allee effect created by predation from zooplankton. In both cases some external event is required to 'kick' the pest population out of a trough of low population growth rate created by predation—in this case excessive nutrient addition or perhaps the loss of macrophytes leading to a decline in the predator (zooplankton) population size. Thus the complex set of multiple-species interactions can to some extent be understood in terms of single species population dynamics (Murdoch *et al.* 2002), giving an insight into the way such ecosystem thresholds arise out of population thresholds.

Another well-known example is that of Caribbean coral reefs, where the ecosystem has switched from coral-dominated to macrophyte-dominated after the loss of herbivores to overfishing (fish) and disease (urchins). In this case it is the macrophytic algae which is the 'pest' species that can be regarded as being released from a predation-driven Allee effect. The alternative state is stable because the algae outcompete coral recruits for space and because many algae species are only palatable to herbivores when small (Knowlton 1992, Hughes 1994, Scheffer *et al.* 2001). Predator-driven Allee effects may likewise underlie the outbreaks of crown-of-thorns starfish on Indo-Pacific reefs, with significant effects at the ecosystem level (Dulvy *et al.* 2004).

6.4. **Allee effects in other sciences**

One of our objectives in this final chapter was to provide some less 'mainstream' material (from an ecologist's point of view at least), to provoke discussion and widen perspectives. This endeavour calls for a glimpse of Allee effects (or Allee-type effects) as seen by other scientific communities. Obviously, we are not going to review analogies of extinction thresholds in quantum physics or organic chemistry (apart from anything, we are entirely unqualified for the task), but without going that far, Allee effects are starting to receive some attention in fields quite foreign to most ecologists.

For example, Allee effects are receiving attention among mathematicians, and both the tools and the perspectives they adopt are somewhat different from the ones we are familiar with. The scope of mathematical publications on Allee effects is wide. Some of the key general topics recently explored in mathematical journals include the following: (i) exploration of 'delay differential equations' in which the change in population density is not only a function of the current density but also of population density at some earlier time; these types of equations may well be good descriptions of population processes such as the time taken to reach maturity or the gestation time (e.g. Gopalsamy and Ladas 1990, Song *et al.* 2004, Sun and Saker 2005a, 2005b); (ii) advanced analysis of models already known from the biological literature (e.g. Jang 2006, Sugie and Kimoto 2006); (iii) in-depth exploration of a general class of Allee effect models (e.g. Gyllenberg *et al.* 1996); or (iv) proving the existence of and finding properties of specific model solutions (e.g. Petrovskii *et al.* 2005, Sun and Saker 2005a, 2005b, Yan *et al.* 2005). These papers are mostly concerned with the exploration of various 'technical' properties of mathematical models involving Allee effects, and are thus of marginal interest for ecology, but nonetheless deserve a brief mention here.

At the other extreme of the so-called 'hard/soft sciences' continuum is the wide domain of social sciences, some of which have also touched on Allee effects, for example in humans. As suggested in our slightly tongue-in-cheek foreword, some aspects of human social life might be driven by extinction thresholds. In fact, demographics of human populations have been shown to demonstrate some similarities to Allee effects. For example, Sardinia in the Middle Ages was characterized by a declining human population. Records from the time show that entire villages emptied as soon as their population dropped below a critical level at which manpower became insufficient to cultivate the land properly; the remaining inhabitants had to leave or face dying of hunger (Day 1975). At some level, the current exodus from rural areas in much of western Europe shares similarities: once a critical number of inhabitants has been reached in a village or a rural area, social infrastructure such as schools, post offices and shops no longer stay cost-effective and have to close, accelerating the population drain from the area.

Still in the social sciences, a recent study has revealed similiarities in patterns between the extinction of biodiversity and extinction of human languages (Sutherland 2003). This study showed that if the same criteria are used, then languages are currently more threatened than birds or mammals, with rare languages being more likely to show evidence of decline than commoner ones (Fig. 6.2). An obvious mechanism suggested by the author is that as languages become rare, they get less attractive for people to learn and use.

Returning closer to the main field of this book (ecology and conservation), one key connection between social sciences and Allee effects has been created by the recent work on the anthropogenic Allee effect (Courchamp *et al.* 2006); see Section 5.3.3. This process has repercussions in psychosociology, economics,

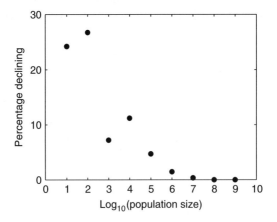

Figure 6.2. Percentage of languages documented as declining. From Sutherland (2003). Copyright: Frans Lanting/Corbis.

environmental law and management sciences, as well as ecology and conservation. Indeed, it could be argued that this process is an economic effect rather than an ecological one. More generally, it is interesting to note how the economic idea of the 'discount rate' leads to ecologically irrational but economically rational harvesting decisions that create a predation (exploitation) driven Allee effect (see Section 5.2.2).

6.5. The future of Allee effects

Without having to go as far as delayed differential equations or medieval Sardinians, there are many areas of research in our discipline in which Allee effect studies could be developed. Based on the content of this book and on an extensive literature search, it appears that ecology and conservation would benefit most from further Allee effect studies in four directions: (i) extending the search for Allee effects into new species, particularly those at risk from climate change, habitat loss, etc.; (ii) research using new model taxa, or at least model species; (iii) research on new types of Allee effects; and (iv) further exploration of the consequences of Allee effects for population and community dynamics.

6.5.1. Allee effects in new species

A search for Allee effects in new species could achieve two main goals. First, demonstrating an Allee effect in a species would obviously be important for conservation and management, either for characterizing likely new threats to its survival or for fighting against long-established ones. Secondly and more generally, a search for Allee effects in new species might help find new Allee effect mechanisms. As an example, starting from the principle that a basic mechanism creating a component Allee effect in survival is environmental conditioning (the attempt of individuals to reduce the area or time exposed to unfavourable conditions, see Section 2.3.1), ecologists might fruitfully consider species which have gregarious behaviours dictated by harsh weather. To survive the cold, mature Emperor penguins *Aptenodytes forsteri* stand in compact huddles (the so-called turtle formation) in large colonial nesting areas during the Antarctic winter. Those on the outside of this formation are more exposed to the cold and tend to shuffle slowly around the edge, giving each bird a turn on the inside and the outside. It is easy to imagine that larger colonies will have relatively fewer individuals on the outside, and that there may be a critical colony size below which there are too few birds for each to warm up long enough in the centre of the formation before their turn outside comes again. It is also known that predators grab individuals at the edge of these formations, and that smaller formations will thus have relatively more vulnerable individuals than larger colonies. As a result, fast churning (i.e. small) formations would contain birds that suffer more from mortality (from the Antarctic winter cold or from predators). This could be tested by looking for relationships between colony size and survival rates (for example). This is highly speculative, as no data have been analysed to test this hypothesis, or, to our knowledge, even collected. We are not penguin experts, but timidly suggest that perhaps Allee effects deserve further investigation in this species, as well as in other species for which similar mechanisms could be envisioned.

Other species that might merit new Allee-effect oriented studies include eusocial species such as ants, termites, bees, wasps, spiders and naked mole rats. The very nature of the social structure of these species make them strong candidates for all sorts of component Allee effect (see Box 2.7), and it is surprising that studies of Allee effects are virtually non-existent in these species (Avilés and Tufiño 1998 being an interesting exception).

Similarly, it would be interesting to consider, from an Allee effect point of view, the enormous pool of obligate associations of all sorts, such as specialist pollinators or symbionts. For example, there are more than one thousand pairs of fig trees and chalcid wasp species, each pair being completely dependent upon each other for the completion of their life cycle, and each living naturally in fairly low densities (e.g. Machado *et al.* 2005), where reproduction-related Allee effects of both have to be overcome for each population to persist.

Other taxa could be investigated simply because they have not been considered much to date. For example, reptiles are poorly represented in the Allee effect literature. Similarly, the American toad is the only potential example of an Allee

Figure 6.3. Emperor penguin chicks huddling to fight cold. It is possible that smaller colonies experience higher instablity rates as individuals are more exposed to extreme conditions. Copyright: Gary Bell/Zefa/Corbis.

effect we have found in an amphibian. Given their recent precipitous decline in numbers (Kiesecker *et al.* 2001), amphibian population dynamics may unfortunately provide fertile hunting grounds for Allee effect researchers in the future.

Of course, we have already made the point that collecting data which demonstrates an Allee effect is not easy, because of the requirement to collect data on individual fitness from small or sparse populations, and across a range of population sizes or densities, as well as because of the risk of confounding variables discussed above. This endeavour therefore has to go hand in hand with development of improved techniques of collecting relevant data on individual fitness in small or sparse populations, as well as techniques for statistical analysis of such data.

In addition, provided adequate data collection is possible, an approach that would really be beneficial for our understanding of Allee effect processes is interspecies comparisons. In particular, a meta-analysis of large sets of time series from populations of many different species, with contrasted life histories is likely to provide vital new information on the traits that are the most likely to make a species sensitive to Allee effects.

This approach might, for example, allow us to test the hypothesis discussed in Chapter 4, that counter-intuitively, naturally large and dense populations are more likely to be sensitive to Allee effects than naturally small or sparse populations. If this is true, krill or locusts or herds of ungulates, for example, could be more sensitive to Allee effects than, say, rare orchids or black-footed cats *Felis nigripes*. The saiga antelope discussed in Section 6.2.3 has an interesting recent history in this regard. At the end of the 1920s, its world population amounted to no more than a few hundreds, due primarily to intensive hunting and to severe 'djusts' (glazed

Figure 6.4. Eusocial insects, such as ants, are strong candidates for different types of component Allee effect.

snow coverings of great persistence; Stoddart 1974). The saiga has a very high reproductive rate, and a combination of several mild winters with a cessation of hunting led to such a dramatic growth that in 25 years, they numbered over 1 million, and 2 million another 20 years later despite a reopening of the commercial hunt. In the past ten years, the global population have dropped from around one million to less than 50 000 (Milner-Gulland *et al.* 2003). The point is, this species seems to have kicked back from a few hundreds to a couple of million in no time, and therefore they seem to have not suffered (overly) from Allee effects. In fact, the above hypothesis might hold for this species. A very high reproductive rate for such a large ungulate would be consistent with an unpredictable environment that probably results in high mortality. It is thus possible that this species could have a natural history of rapid fluctuations, being sensitive to mortality causes such as djusts, but bouncing back quickly. According to this hypothesis, this species would thus be 'freed' from any Allee effect. This would have two implications for conservation: (i) saiga antelope may be especially sensitive to hunting, (ii) hunting bans could allow them to recover quickly.

6.5.2. **Research using model taxa**

The second possible avenue of research is based on taxa or species that could prove excellent models not only for the species in question, but for the process itself. In-depth studies using a well-chosen model organism could be a powerful complement to theoretical studies in providing insights into poorly understood facets of the mechanisms for, and consequences of, Allee effects.

Plants provide useful model species, since they are sessile and their populations can generally be more easily manipulated for dynamical or genetic studies (see Le Cadre *et al.*, in review). Plants may also have Allee effect mechanisms which are less well-documented in animals, such as collective modification of the environment or genetic Allee effects. Other possibilities include protists, bacteria, fungi or algae (Le Cadre *et al.*, in review).

As well as new species models, studies on Allee effect could focus on system models that can overcome the impracticability and unethicality of experimenting on small populations. Two such systems are biological control and biological invasions, the former amounting in a sense to 'programmed experiments' while the latter can be considered as 'natural experiments'. Both areas provide excellent opportunities to study Allee effects (e.g. Taylor and Hastings 2005, Drake and Lodge 2006, Fauvergue *et al.* 2007).

6.5.3. **Research on new types of Allee effects**

The third direction that would be interesting for future research on Allee effects is to consider new types of positive density dependence. We can give three brief examples here, although of course these are not all inclusive—there are likely to be many more. The first one pertains to the anthropogenic Allee effect described in Section 5.2.3. This new, artificial process has just been described and it opens many research possibilities. For example, it would be interesting to demonstrate, with hard data, whether it occurs in specific activities like luxury markets, ecotourism or traditional medicine, and to identify in more detail the species concerned by this threat. There are a great many issues that are related to this hypothesis, especially if one remembers that conservation sciences encompass sociology, sociopsychology, economics and the like and that only the biological aspects of conservation biology has been studied so far in this context.

The second category of new perspectives concerns the overlooked idea of Allee effects at intermediate population sizes or densities. A hump-shaped relationship between the per capita population growth rate and population size or density is the most common representation of demographic Allee effects (Fig.1.5). The definition of component Allee effects we adopt in this book (which stems from that suggested by Stephens *et al.* 1999) admits, however, one more scenario— positively density-dependent and negatively density-dependent mechanisms in a population may combine to produce a wave-like form of the per capita population growth rate (Fig. 6.5). One may then speak of demographic Allee effects at intermediate population sizes or densities (Stephens *et al.* 1999), the phenomenon that hitherto gained negligible attention among modellers. This scenario

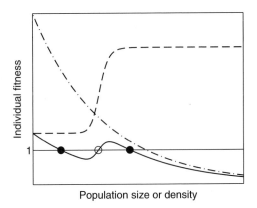

Figure 6.5. Demographic Allee effects at intermediate population sizes or densities. Total individual fitness (thick solid line) is here illustrated as a product of components of fitness due to all positively density-dependent mechanisms (dashed line) and components of fitness due to all negatively density-dependent mechanisms (dash-dot line). The thin solid line denotes the fitness value at which the population is just replacing itself. In this case, the population has three interior equilibria, two of which are locally stable (full dots) and one unstable (open dot); the origin is also unstable here so that any small or sparse population increases and attains the lower stable equilibrium (see also Stephens *et al.* 1999).

may result in up to three interior steady states, two of which are (locally) stable and one unstable; the origin—i.e. extinction—is also unstable in this case.

Up to now, such a wave-like form has been shown to arise in two ways. Firstly, it can occur if the cues used by individuals to assess their fitness in a habitat do not reflect actual habitat quality (Kokko and Sutherland 2001). Unfortunately, no underlying component Allee effect has been suggested to generate this form. Secondly, and probably more commonly, it can arise in a predator–prey system in which prey grows logistically in the absence of predators (i.e. no Allee effect intrinsic to prey) and where generalist predators with no numerical response feed on prey via a type III functional response (see Fig. 3.3). Type III functional responses do not generate the same predation-driven Allee effect as type II responses (Section 3.2.1) but still generate multiple stable states at which both prey and predators may persist, precisely as in Figure 6.5—yet we cannot speak of demographic Allee effect at intermediate population densities here. The lower stable prey equilibrium is called the 'predator pit' in this context—once prey populations become rare they can no longer attain their carrying capacity. Only a prey population augmentation to levels above the unstable threshold can recover the originally high prey densities (May 1977b). Allee effects of this kind might be responsible for a lack of recovery (but also lack of any further decline) of many fisheries following a ban on further exploitation (see Section 5.2, Fig. 5.6).

Finally, the third category of new types of Allee effect is well illustrated by a recent study on cultural transmission in birds (Laiolo and Tella 2007). In a study of variation of non-genetically transmitted traits—song and call repertoires—the authors found an interesting Allee-like effect. In Dupont's lark (*Chersophilus duponti*), like in many other bird species, social learning leads to the acquisition of a diversity of songs. When there are more males, each male has a higher diversity of songs, whereas with three males or fewer, the repertoire is much reduced, when they sing at all. This study showed that when the population decreased, individual and population song repertoires significantly declined in variety (Fig. 6.6) and cultural transmission to juveniles was impoverished: they learned poorer songs. This study also showed that this led to patch disappearance in some cases: acoustic repertoire reduction could be the prelude to population extinction (Laiolo and Tella 2007). We speculate that because male quality is advertised by their songs and selected by females, having a poorer song equates to becoming less attractive, which in turn leads to fewer mating opportunities. As female fitness is increased if she has sons that can learn a high diversity of songs (i.e. when there are more males to learn from), females in large populations have a higher fitness: they choose better quality males, and their sons learn more diverse songs and will be more likely to be selected in turn. This peculiar process therefore really fits with the definition of Allee effects. It is quite likely that other, unusual, Allee-like processes remain to be discovered.

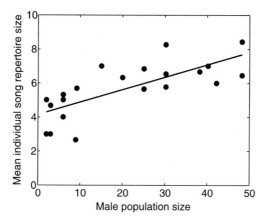

Figure 6.6. Positive relationship between mean song repertoire size of Dupont's lark individuals and male population size. After Laiolo and Tella (2007).

6.5.4. **Consequences of Allee effects for population and community dynamics**

The future of Allee effects awaits us not only on the empirical side of the topic, but also on its theoretical side. There is still much to reveal by modelling, and we discuss here only two promising avenues in some detail.

Firstly, we can learn much about effects of positive density dependence on population dynamics by recognizing and modelling impacts of multiple Allee effects, i.e. in two or more fitness components. We showed in Chapter 2 (Table 2.4) that multiple Allee effects might not be uncommon and in Section 3.2.2 that consequences of multiple Allee effects for population dynamics cannot be disregarded in the management of threatened, pest, or exploited populations. Apart from a need to demonstrate the prevalence and strength of these effects empirically, population models with two or more component Allee effects need to be developed and analysed. This covers not only general-purpose (strategic) models aimed at understanding the general consequences of multiple Allee effects for population dynamics, but also specific (tactic) models of the species listed in Table 2.4. This endeavour is especially important in systems where human activity has a significant role (this probably applies to most systems demonstrating demographic Allee effects; see Table 2.2), if sustainable development and biodiversity conservation are to remain our primary goals.

The second avenue which holds much promise for further theoretical research is to explore the implications of Allee effects for multiple-species systems. As we made clear in Section 3.6, research on impacts of Allee effects in predator–prey systems is well covered, but studies on other types of interspecific interactions (host–parasite relationships, competition, mutualisms) remain relatively rare. Data are now emerging on Allee effects in pathogens or parasites which could stimulate development and analysis of new host-parasite models with Allee effects. Models of mutualisms could go hand in hand with the empirical research on mutualistic interactions suggested above. Predator–prey theory has recently witnessed the discovery of emergent Allee effects among top predators which are size- or stage-selective with respect to their prey (Section 3.6.1) and it is possible that new emergent Allee effects will appear in the future if other models of interspecific interactions also become state-structured. Finally, studies on the implications of Allee effects for the dynamics and structure of larger and more complex food webs are virtually absent and therefore this is an area of a promising and rich research.

If all these possible avenues for research seem important, future progress in understanding the dynamics of small populations will most likely be achieved through a creative combination of all these different scientific approaches.

6.6. **Farewell remarks**

We hope we have convinced you that Allee effects are everywhere, from the survival of the goldfish in your living room tank to the abrupt alpine tree line you can see from the window behind. It is also our hope that you are persuaded that many, if not most, species are confronted with Allee effects, either directly or through species they interact with, and that Allee effects can thus have significant repercussions for the ecology and evolution of species, communities and ecosystems. Because of both their ubiquity and their importance, we finally hope that Allee effects will now be more systematically taken into consideration in the management of populations, either for their sustainable exploitation or for their effective protection. And lastly, we hope that you will convey this message to others. Like many things, information is subject to extinction thresholds: if you keep to yourself the results of this synthesis, it is more likely to become extinct in the end. And you wouldn't want to be the cause of extinction of the Allee effect concept, would you?

References

Abramsky Z., Rosenzweig M. and Subach A. 1997. Gerbils under threat of owl predation: isoclines and isodars. *Oikos*, **78**, 81–90.

Adam M., Sibert J., Itano D. and Holland K. 2003. Dynamics of bigeye (*Thunnus obesis*) and yellowfin (*T. albacares*) tuna in Hawaii's pelagic fisheries: analysis of tagging data with a bulk transfer model incorporating size-specific attrition. *Fisheries Bulletin*, **101**, 215–28.

Adams E.S. and Tschinkel W.R. 2001. Mechanisms of population regulation in the fire ant *Solenopsis invicta*: an experimental study. *Journal of Animal Ecology*, **70**, 355–69.

Agren J. 1996. Population size, pollinator limitation and seed set in the self-incompatible herb *Lythrum salicaria*. *Ecology*, **77**, 1779–90.

Alftine K. and Malanson G. 2004. Directional positive feedback and pattern at an alpine tree line. *Journal of Vegetation Science*, **15**, 3–12.

Al-Husaini M., Al-Baz A., Al-Ayoub S., Safar S., Al-Wazan Z. and Al-Jazzaf S. 2002. Age, growth, mortality, and yield-per-recruit for nagroor, *Pomadasys kakaan*, in Kuwait's waters. *Fisheries Research*, **59**, 101–115.

Allam S. 2003. Growth, mortality and yield per pecruit of bogue, *Boops boops* (L.), from the Egyptian Mediterranean waters off Alexandria. *Mediterranean Marine Science*, **4**, 87–96.

Allee W.C. 1931. *Animal aggregations, a study in general sociology*. Chicago: University of Chicago Press.

Allee W.C. 1941. *The Social Life of Animals*. 2nd or 3rd. William Heineman Ltd, London and Toronto.

Allee W.C., Emerson A., Park O., Park T. and Schmidt K. 1949. *Principles of Animal Ecology*. WB Saunders Company, Philadelphia.

Allee W.C. and Bowen E. 1932. Studies in animal aggregations: mass protection against colloidal silver among goldfishes. *Journal of Experimental Zoology*, **61**, 185–207.

Allee W.C. and Rosenthal G.M.J. 1949. Group survival value for *Philodina roseola*, a rotifer. *Ecology*, **30**, 395–7.

Allee W.C. and Wilder J. 1938. Group protection for *Euplanaria dorotocephala* from ultraviolet radation. *Physiological Zoology*, **12**, 110–35.

Allen L.J.S., Fagan J.F., Högnäs G., and Fagerholm H. 2005. Population extinction in discrete-time stochastic population models with an Allee effect. *Journal of Difference Equations and Applications*, **11**, 273–93.

Allendorf F.W. and Luikart G. 2006. Conservation and the genetics of populations. Blackwell Publishing Limited.

Allison T. 1990. Pollen production and plant density affect pollination and seed production in *Taxus canadensis*. *Ecology*, **71**, 516–22.

Almeida R.C., Delphim S.A., and da S. Costa M.I. 2006. A numerical model to solve single-species invasion problems with Allee effects. *Ecological Modelling*, **192**, 601–617.

Alvarez L.H.R. 1998. Optimal harvesting under stochastic fluctuations and critical depensation. *Mathematical Biosciences*, **152**, 63–85.

Ama-Abasi D., Holzloehner S., and Enin U. 2004. The dynamics of the exploited population of *Ethmalosa fimbriata* (Bowdich, 1825, Clupeidae) in the Cross River Estuary and adjacent Gulf of Guinea. *Fisheries Research*, **68**, 225–35.

Amarasekare P. (1998a). Allee effects in metapopulation dynamics. *American Naturalist*, **152**, 298–302.

Amarasekare P. (1998b). Interactions between local dynamics and dispersal: insights from single species models. *Theoretical Population Biology*, **53**, 44–59.

Amarasekare P. 2004. Spatial dynamics of mutualistic interactions. *Journal of Animal Ecology*, **73**, 128–42.

Anderson J. and Rose G. 2001. Offshore spawning and year-class strength of northern cod (2J3KL) during the fishing moratorium, 1994–96. *Canadian Journal of Fisheries and Aquatic Science*, **59**, 1386–94.

Anderson R.M. and May R.M. 1979. Population biology of infectious diseases: part II. *Nature*, **280**, 455–61.

Angulo E., Rasmussen G., Macdonald D.W. and Courchamp F. 2008. Demographic and component Allee effects in the African wild dog, *hycaon pictus*. *Conservation Biology*, in press.

Angulo E., Roemer G.W., Berec L., Gascoigne J., and Courchamp F. 2007. Double Allee effects and extinction in the island fox. *Conservation Biology*.

Apollonio M., Bassano B., and Mustoni A. 2003. Behavioral aspects of conservation and management of European mammals. In M. Festa-Bianchet and M. Apollonio, eds. *Animal behavior and wildlife conservation*, pp. 157–70. Island Press, Washington D.C.

Appeldoorn R.S. 1988. Fishing pressure and reproductive potential in strombid conchs: is there a critical stock density for reproduction? *Memoria de la Sociedad de Ciencias Naturales La Salle*, **48**, 275–88.

Aronson R. and Precht W. 2001. White-band disease and the changing face of Caribbean coral reefs. *Hydrobiologia*, **460**, 25–38.

Arrontes J. 2005. A model for range expansion of coastal algal species with different dispersal strategies: the case of *Fucus serratus* in northern Spain. *Marine Ecology Progress Series*, **295**, 57–68.

Ashih A.C. and Wilson W.G. 2001. Two-sex population dynamics in space: effects of gestation time on persistence. *Theoretical Population Biology*, **60**, 93–106.

Ashman T.L., Knight T.M., Steets J.A., *et al.* 2004. Pollen limitation of plant reproduction: ecological and evolutionary causes and consequences. *Ecology*, **85**, 2408–421.

Asmussen M.A. 1979. Density-dependent selection II. The Allee effect. *American Naturalist*, **114**, 796–809.

Aukema B. and Raffa K. 2004. Does aggregation benefit bark beetles by diluting predation? Links between a group-colonisation strategy and the absence of emergent multiple predator effects. *Ecological Entomology*, **29**, 129–38.

Avilés L. 1999. Cooperation and non-linear dynamics: an ecological perspective on the evolution of sociality. *Evolutionary Ecology Research*, **1**, 459–77.

Avilés L. and Tufiño P. 1998. Colony size and individual fitness in the social spider *Anelosimus eximius*. *American Naturalist*, **152**, 403–18.

Babcock R. and Keesing J. 1999. Fertilisation biology of the abalone *Haliotis laevigata*: laboratory and field studies. *Canadian Journal of Fisheries and Aquatic Science*, **56**, 1668–78.

Banks E.M. 1985. Warder Clyde Allee and the Chicago School of Animal Behavior. *Journal of the History of Behavioral Sciences*, **21**, 345–53.

Barclay H. and Mackauer M. 1980a. The sterile insect release method for pest control: a density-dependent model. *Environmental Entomology*, **9**, 810–17.

Barclay H. and Mackauer M. 1980b. Effects of sterile insect releases on a population under predation or parasitism. *Researches on Population Ecology*, **22**, 136–46.

Barrowman N.J., Myers R.A., Hilborn R., Kehler D.G., and Field C.A. 2003. The variability among populations of coho salmon in the maximum reproductive rate and depensation. *Ecological Applications*, **13**, 784–93.

Baumhover A., Graham A., Bitter B., Hopkins D., New W., Dudley F., and Buchland R. 1955. Screw-worm control through release of sterilised flies. *Journal of Economic Entomology*, **48**, 462–6.

Beadman H.A., Kaiser M.J., Galanidi M., Shucksmith R., and Willows R.I. 2004. Changes in species richness with stocking density of marine bivalves. *Journal of Applied Ecology*, **41**, 464–75.

Begon M., Harper J., and Townsend C. 1996. *Ecology: Individuals, Populations and Communities*. Blackwell, Oxford.

Becheikh S., Michaud M., Thomas F., Raibaut A., and Renaud F. 1998. Roles of resource and partner availability in sex determination in a parasitic copepod. *Proceedings of the Royal Society of London B*, **265**, 1153–56.

Berec L. 2002. Techniques of spatially explicit individual-based models: construction, simulation, and mean-field analysis. *Ecological Modelling*, **150**, 55–81.

Berec L. and Boukal D.S. 2004. Implications of mate search, mate choice and divorce rate for population dynamics of sexually reproducing species. *Oikos*, **104**, 122–32.

Berec L., Angulo E., and Courchamp F. 2007. Multiple Allee effects and population management. *Trends in Ecology & Evolution*, **22**, 185–91.

Berec L., Boukal D.S., and Berec M. 2001. Linking the Allee effect, sexual reproduction, and temperature-dependent sex determination via spatial dynamics. *American Naturalist*, **157**, 217–30.

Berg A., Lindberg T., and Källebrink K.G. 1992. Hatching success of lapwings on farmland: differences between habitats and colonies of different sizes. *Journal of Animal Ecology*, **61**, 469–76.

Berg C. and Olsen D. 1989. Conservation and management of queen conch (*Strombus gigas*) fisheries in the Caribbean. In J. Caddy, ed. *Marine invertebrate fisheries: their assessment and management*, pp. 421–42. Wiley and Sons, New York.

Berggren A. 2001. Colonisation success is Roesel's bush-cricket *Metrioptera roeseli*: the effects of propagule size. *Ecology*, **82**, 274–80.

Berryman A.A. 2003. On principles, laws and theory in population ecology. *Oikos*, **103**, 695–701.

Bertness M. and Grosholz E. 1985. Population dynamics of the ribbed mussel, *Geukensia demissa*: the costs and benefits of a clumped distribution. *Oecologia*, **67**, 192–204.

Beshers S.N. and Traniello J.F.A. 1994. The adaptiveness of worker demography in the attine ant, *Trachymyrmex septentrionalis*. *Ecology*, **75**, 763–75.

Bessa-Gomes C., Legendre S., and Clobert J. 2004. Allee effects, mating systems and the extinction risk in populations with two sexes. *Ecology Letters*, **7**, 802–12.

Billick I., Wiernasz D.C., and Cole B.J. 2001. Recruitment in the harvester ant, *Pogonomyrmex occidentalis*: effects of experimental removal. *Oecologia*, **129**, 228–33.

Boland C., Heinsohn R., and Cockburn A. 1997. Deception by helpers in cooperatively breeding white-winged choughs and its experimental manipulation. *Behavioral Ecology and Sociobiology*, **41**, 251–6.

Bond W. 1994. Do mutualisms matter? Assessing the impact of pollinator and disperser disruption on plant extinction. *Philosophical Transactions of the Royal Society of London B*, **344**, 83–90.

Boomsma J.J., van der Lee G.A., and van der Have T.M. 1982. On the production ecology of *Lasius niger* (Hymenoptera: Formicidae) in successive coastal dune valleys. *Journal of Animal Ecology*, **51**, 975–91.

Borowicz V. and Juliano S. 1986. Inverse density-dependent parasitism of *Cornus amomum* fruit by *Rhagoletis cornivora*. *Ecology*, **67**, 639–43.

Botham M.S. and Krause J. 2005. Shoals receive more attacks from the wolf-fish (*Hoplias malabaricus* Bloch, 1794). *Ethology*, **111**, 881–90.

Boukal D.S. and Berec L. 2002. Single-species models of the Allee effect: extinction boundaries, sex ratios and mate encounters. *Journal of Theoretical Biology*, **218**, 375–94.

Boukal D.S., Sabelis M.W., and Berec L. 2007. How predator functional responses and Allee effects in prey affect the paradox of enrichment and population collapses. *Theoretical Population Biology*.

Bourke A.F.G., van der Have T.M., and Franks N.R. 1988. Sex ratio determination and worker reproduction in the slave-making ant *Harpagoxenus sublaevis*. *Behavioral Ecology and Sociobiology*, **23**, 233–45.

Boveng P., Hiruki L., Schwartz M., and Bengtson J. 1998. Population growth of antarctic fur seals: limitation by a top predator, the leopard seal? *Ecology*, **79**, 2863–77.

Boyce M. 1992. Population viability analysis. *Annual Review of Ecology and Systematics*, **23**, 481–506.

Bradford E. and Philip J.R. 1970. Note on asocial populations dispersing in two dimensions. *Journal of Theoretical Biology*, **29**, 27–33.

Brassil C.E. 2001. Mean time to extinction of a metapopulation with an Allee effect. *Ecological Modelling*, **143**, 9–16.

Breitenmoser U., Breitenmoser-Wursten C., Carbyn L., and Funk S. 2001. Assessment of carnivore reintroductions. In J. Gittleman, S. Funk, D. Macdonald and R. Wayne, eds. *Carnivore conservation*, pp. 490–530. Cambridge University Press, Cambridge.

Brian M.V. and Elmes G.W. 1974. Production by the ant *Tetramorium caespitum* in a southern English heath. *Journal of Animal Ecology*, **43**, 889–903.

Brockett B. and Hassall M. 2005. The existence of an Allee effect in populations of *Porcellio scaber* (Isopoda: Oniscidea). *European Journal of Soil Biology*, **41**, 123–7.

Brook B., O'Grady J., Chapman A., Burgman M., Akcakaya H., and Frankham R. 2000. Predictive accuracy of population viability analysis in conservation biology. *Nature*, **404**, 385–7.

Brook B.W., Traill L.W., and Bradshaw C.J.A. 2006. Minimum viable populations and global extinction risk are unrelated. *Ecology Letters*, **9**, 375–82.

Brown C. and Brown M. 2002. Spleen volume varies with colony size and parasite load in a colonial bird. *Proceedings of the Royal Society of London B*, **269**, 1367–73.

Brown C. and Brown M. 2003. Testis size increases with colony size in cliff swallows. *Behavioral Ecology*, **14**, 569–75.

Brown C. and Brown M. 2004. Group size and ectoparasitism affect daily survival probability in a colonial bird. *Behavioural Ecology and Sociobiology*, **56**, 498–511.

Brown C., Covas R., Anderson M., and Brown M. 2003. Multistate estimates of survival and movement in relation to colony size in the sociable weaver. *Behavioral Ecology*, **14**, 463–71.

Brown C., Stutchbury B., and Walsh P. 1990. Choice of colony size in birds. *Trends in Ecology & Evolution*, **5**, 398–403.

Brown D.H., Ferris H., Fu S., and Plant R. 2004. Modeling direct positive feedback between predators and prey. *Theoretical Population Biology*, **65**, 143–52.

Bruno J.F., Stachowicz J.J. and Bertness M.D. 2003. Inclusion of facilitation into ecological theory. *Trends in Ecology & Evolution*, **18**, 119–25.

Burd M. 1994. Bateman's principle and plant reproduction: the role of pollen limitation in fruit and seed set. *The Botanical Review*, **60**, 83–139.

Burgman M.A., Ferson S., and Akçakaya H.R. 1993. *Risk assessment in conservation biology*. Chapman & Hall, London.

Butterworth D., Korrubel J., and Punt A. 2002. What is needed to make a simple density-dependent response population model consistent with data for eastern North Pacific gray whales? *Journal of Cetacean Resource Management*, **4**, 63–76.

Buza L., Young A., and Thrall P. 2000. Genetic erosion, inbreeding and reduced fitness in fragmented populations of the endangered tetraploid pea *Swainsona recta*. *Biological Conservation*, **93**, 177–86.

Byers D.L. 1995. Pollen quantity and quality as explanations for low seed set in small populations exemplified by *Eupatorium* (Asteraceae). *American Journal of Botany*, **82**, 1000–6.

Byers D.L. and Meagher T.R. 1992. Mate availability in small populations of plant species with homomorphic sporophytic self-incompatibility. *Heredity*, **68**, 353–9.

Byers D.L. and Waller D.M. 1999. Do plant populations purge their genetic load? Effects of population size and mating history on inbreeding depression. *Annual Review of Ecology and Systematics*, **30**, 479–513.

Calabrese J.M. and Fagan W.F. 2004. Lost in time, lonely, and single: reproductive asynchrony and the Allee effect. *American Naturalist*, **164**, 25–37.

Calvert W., Hedrick L., and Brower L. 1979. Mortality of the monarch butterfly (*Danaus plexippus* L.): avian predation at five overwintering sites in Mexico. *Science*, **204**, 847–51.

Cappuccino N. 2004. Allee effect in an invasive alien plant, pale swallow-wort Vincetoxicum rossicum (Asclepiadaceae). Oikos, **106**, 3–8.

Carpenter S.R. 2003. *Regime shifts in lake ecosystems: pattern and variation*. Ecology Institute, Oldendorf/Luhe, Germany.

Carver A.M., Wolcott T.G., Wolcott D.L. and Hines A.H. 2005. Unnatural selection: effects of a male-focused size-selective fishery on reproductive potential of a blue crab population. *Journal of Experimental Marine Biology and Ecology*, **319**, 29–41.

Case T.J. 2000. *An illustrated guide to theoretical ecology*. Oxford University Press, New York and Oxford.

Caswell H. 2001. *Matrix population models. Construction, analysis, and interpretation.* Second Edition. Sinauer Associates, Sunderland, Massachusetts.

Caughley G. 1994. Directions in conservation biology. *Journal of Animal Ecology*, **63**, 215–44.

Charlesworth D. and Charlesworth B. 1987. Inbreeding depression and its evolutionary consequences. *Annual Review of Ecology and Systematics*, **18**, 237–68.

Chen D.G., Irvine J.R., and Cass A.J. 2002. Incorporating Allee effects in fish stock-recruitment models and applications for determining reference points. *Canadian Journal of Fisheries and Aquatic Sciences*, **59**, 242–9.

Chen Z., Qiu Y., and Huang Z. 2005. Estimation of growth and mortality parameters of *Argyrosomus argentatus* in northern South China Sea. *Ying yong sheng tai xue bao = The Journal of Applied Ecology*, **16**, 712–16.

Cheptou P.O. 2004. Allee effect and self-fertilization in hermaphrodites: reproductive assurance in demographically stable populations. *Evolution*, **58**, 2613–21.

Chidambaram J., Lee D., Porco T., and Lietman T. 2005. Mass antibiotics for trachoma and the Allee effect. *Lancet Infectious Diseases*, **5**, 194–6.

Clark C. 1976. *Mathematical bioeconomics: the optimal management of renewable resources.* John Wiley, New York.

Clutton-Brock T., Gaynor D., McIlrath G., *et al.* 1999. Predation, group size and mortality in a cooperative mongoose, *Suricata suricatta. Journal of Animal Ecology*, **68**, 672–83.

Cole C.J. 1984. Unisexual lizards. *Scientific American*, **250**, 94–100.

Cole B.J. and Wiernasz D.C. 2000. Colony size and reproduction in the western harvester ant, *Pogonomyrmex occidentalis. Insectes Sociaux*, **47**, 249–55.

Conover D.O. and Munch S.B. 2002. Sustaining fisheries yields over evolutionary time scales. *Science*, **297**, 94–6.

Cook W., Holt R., and Yao J. 2001. Spatial variability in oviposition damage by periodical cicadas in a fragmented landscape. *Oecologia*, **127,** 51–61.

Cornell S.J. and Isham V.S. 2004. Ultimate extinction of the promiscuous bisexual Galton–Watson metapopulation. *Australian & New Zealand Journal of Statistics*, **46**, 87–98.

Cornell S.J., Isham V.S., and Grenfell B.T. 2004. Stochastic and spatial dynamics of nematode parasites in farmed ruminants. *Proceedings of the Royal Society of London* Series B-Biological Sciences, **271**, 1243–50.

Cosner C., DeAngelis D.L., Ault J.S., and Olson D.B. 1999. Effects of spatial grouping on the functional response of predators. *Theoretical Population Biology*, **56**, 65–75.

Coulson T., Mace G., Hudson E., and Possingham H. 2001. The use and abuse of population viability analysis. *Trends in Ecology & Evolution*, **16,** 219–21.

Courchamp F. and Cornell S.J. 2000. Virus-vectored immunocontraception to control feral cats on islands: a mathematical model. *Journal of Applied Ecology*, **37**, 903–13.

Courchamp F. and Macdonald D.W. 2001. Crucial importance of pack size in the African wild dog Lycaon pictus. *Animal Conservation*, **4**, 169–74.

Courchamp F., Angulo E., Rivalan P., *et al.* 2006. Rarity value and species extinction: the anthropogenic Allee effect. *Plos Biology*, **4,** 2405–10.

Courchamp F., Clutton-Brock T.H., and Grenfell B.T. 1999a. Inverse density dependence and the Allee effect. *Trends in Ecology & Evolution*, **14**, 405–10.

Courchamp F., Grenfell B.T. and Clutton-Brock T.H. 1999b. Population dynamics of obligate cooperators. *Proceedings of the Royal Society of London Series B-Biological Sciences*, **266**, 557–63.

Courchamp F., Clutton-Brock T.H., and Grenfell B.T. 2000a. Multipack dynamics and the Allee effect in the African wild dog, *Lycaon pictus. Animal Conservation*, **3**, 277–85.

Courchamp F., Grenfell B.T., and Clutton-Brock T.H. 2000b. Impact of natural enemies on obligately cooperative breeders. *Oikos*, **91**, 311–22.

Courchamp F., Rasmussen G., and Macdonald D. 2002. Small pack size imposes a trade-off between hunting and pup-guarding in the painted hunting dog *Lycaon pictus. Behavioral Ecology*, **13,** 20–7.

Courchamp F., Woodroffe R., and Roemer G. 2003. Removing protected populations to save endangered species. *Science*, **302**, 1532.

Crawley M.J. 1983. *Herbivory: the dynamics of animal-plant interactions.* Blackwell Scientific Publications, Oxford.

Creel S., McNutt J., and Mills M. 2004. Demography and Population Dynamics of African Wild Dogs in Three Critical Populations. In D. Macdonald and C. Sillero-Zubiri, eds. *Biology and Conservation of Wild Canids*, pp. 337–50. Oxford University Press, Oxford.

Crews D. and Fitzgerald K. 1980. 'Sexual' behaviour in parthenogenetic lizards (*Cnemidophorus*). *Proceedings of the National Academy of Science of the United States of America*, **77**, 499–502.

Crews D., Grassman M., and Lindzey J. 1986. Behavioural facilitation of reproduction in sexual and unisexual whiptail lizards. *Proceedings of the National Academy of Sciences of the United States of America*, **83**, 9547–50.

Crews D., Teramoto L., and Carson H. 1984. Behavioural facilitation of reproduction in sexual and parthenogenetic *Drosophila*. *Science*, **227**, 77–8.

Crone E.E., Polansky L., and Lesica P. 2005. Empirical models of pollen limitation, resource acquisition, and mast seeding by a bee-pollinated wildflower. *American Naturalist*, **166**, 396–408.

Cushing J.M. 1994. Oscillations in age-structured population models with an Allee effect. *Journal of Computational and Applied Mathematics*, **52**, 71–80.

Cushing J.M., Costantino R.F., Dennis B., Desharnais R.A., and Henson S.M. 2003. *Chaos in ecology. Experimental nonlinear dynamics.* Academic Press, San Diego.

Cuthbert R. 2002. The role of introduced mammals and inverse density-dependent predation in the conservation of Hutton's shearwater. *Biological Conservation*, **108**, 69–78.

Danchin E. and Wagner R. 1997. The evolution of coloniality: the emergence of new perspectives. *Trends in Ecology & Evolution*, **12**, 342–6.

Darvill B., Ellis J.S., Lye C.G., and Goulson D. 2005. Population structure and inbreeding in a rare and declining bumblebee, *Bombus muscorum* (Hymenoptera: Apidae). *Molecular Ecology*, **15**, 601–11.

Davis H.G., Taylor C.M., Lambrinos J.G., and Strong D.R. 2004. Pollen limitation causes an Allee effect in a wind-pollinated invasive grass (*Spartina alterniflora*). *Proceedings of the National Academy of Sciences of the United States of America*, **101**, 13804–7.

Day J. 1975. Malthus contradicted—chronic underpopulation and demographic catastrophe in Sardinia during Early Middle-Ages. *Annales: Economies, Societes, Civilisations*, **30**, 684–702.

de Lange P.J. and Norton D.A. 2004. The ecology and conservation of *Kunzea sinclairii* (Myrtaceae), a naturally rare plant of rhyolitic rock outcrops. *Biological Conservation*, **117**, 49–59.

de Roos A.M. and Persson L. 2002. Size-dependent life-history traits promote catastrophic collapses of top predators. *Proceedings of the National Academy of Sciences of the United States of America*, **99**, 12907–12.

de Roos A.M., Persson L., and Thieme H.R. 2003. Emergent Allee effects in top predators feeding on structured prey populations. *Proceedings of the Royal Society of London Series B-Biological Sciences*, **270**, 611–18.

DeMauro M.M. 1993. Relationship of breeding system to rarity in the lakeside daisy (*Hymenoxys acaulis var. glabra*). *Conservation Biology*, **7**, 542–50.

Dennis B. 1989. Allee effects: population growth, critical density and the chance of extinction. *Natural Resource Modeling*, **3**, 481–538.

Dennis B. 2002. Allee effects in stochastic populations. *Oikos*, **96**, 389–401.

Deredec A. and Courchamp F. 2003. Extinction thresholds in host-parasite dynamics. *Annales Zoologici Fennici*, **40**, 115–30.

Deredec A. and Courchamp F. 2006. Combined impacts of Allee effects and parasitism. *Oikos*, **112**, 667–79.

Deredec A. and Courchamp F. 2007. Importance of the Allee effect for reintroductions. *Ecoscience*, in press.

Devi K.U. and Rao C.U.M. 2006. Allee effect in the infection dynamics of the entomopathogenic fungus *Beauveria bassiana* (Bals) Vuill. on the beetle, *Mylabris pustulata*. *Mycopathologia*, **161**, 385–94.

DeVito J. 2003. Metamorphic synchrony and aggregation as antipredator responses in American toads. *Oikos*, **103**, 75–80.

Dickman C. 1992. Commensal and mutualistic interactions among terrestrial vertebrates. *Trends in Ecology & Evolution*, **7,** 194–7.

Dobson A.P. and Lyles A.M. 1989. The population dynamics and conservation of primate populations. *Conservation Biology*, **3**, 362–80.

Dornelas M., Connolly S.R., and Hughes T.P. 2006. Coral reef diversity refutes the neutral theory of biodiversity. *Nature*, **440**, 80–2.

Drake J.M. 2004. Allee effects and the risk of biological invasion. *Risk Analysis*, **24**, 795–802.

Drake J.M. and Lodge D.M. 2006. Allee effects, propagule pressure and the probability of establishment: risk analysis for biological invasions. *Biological Invasions*, **8**, 365–75.

Dubois F., Cézilly F., and Pagel M. 1998. Mate fidelity and coloniality in waterbirds: a comparative analysis. *Oecologia*, **116**, 433–40.

Dulvy N.K., Freckleton R.P., and Polunin N.V.C. 2004. Coral reef cascades and the indirect effects of predator removal by exploitation. *Ecology Letters*, **7**, 410–16.

Ebert D., Zschokke-Rohringer C.D., and Carius H.J. 2000. Dose effects and density-dependent regulation of two microparasites of *Daphnia magna*. *Oecologia*, **122**, 200–9.

Eckert C.G. 2001. The loss of sex in clonal plants. *Evolutionary Ecology*, **15**, 501–20.

Eckrich C.E. and Owens D.W. 1995. Solitary versus arribada nesting in the olive ridley sea turtles (*Lepidochelys olivacea*): a test of the predator-satiation hypothesis. *Herpetologica*, **51**, 349–54.

Edelstein-Keshet L. 1988. *Mathematical models in biology*. McGraw-Hill.

Eggleston D.B., Johnson E.G., Kellison G.T., and Nadeau D.A. 2003. Intense removal and non-saturating functional responses by recreational divers on spiny lobster *Panulirus argus*. *Marine Ecology-Progress Series*, **257**, 197–207.

Ehrlen J. 1992. Proximate limits to seed production in a herbaceous perennial legume, *Lathyrus vernus*. *Ecology*, **73**, 1820–31.

Ehrlen J. and Eriksson O. 1995. Pollen limitation and population growth in a herbaceous perennial legume. *Ecology*, **76,** 652–6.

Elliott G.P., Merton D.V., and Jansen P.W. 2001. Intensive management of a critically endangered species: the kakapo. *Biological Conservation*, **99**, 121–33.

Ellstrand N.C. and Elam D.R. 1993. Population genetic consequences of small population size: implications for plant conservation. *Annual Review of Ecology and Systematics*, **24**, 217–42.

Elmes G.W. 1987. Temporal variation in colony populations of the ant *Myrmica sulcinodis*. II. Sexual production and sex ratios. *Journal of Animal Ecology*, **56**, 573–83.

Elmes G.W. and Wardlaw J.C. 1982. A population study of the ants *Myrmica sabuleti* and *Myrmica scabrinodis*, living at two sites in the south of England. I. A comparison of colony populations. *Journal of Animal Ecology*, **51**, 651–64.

Emlen S. 1997. Predicting family dynamics in social vertebrates. In J.R. Krebs and N.B. Davies, eds. *Behavioural ecology: an evolutionary approach*, 4th edition, pp. 228–52. Blackwell Scientific, Oxford.

Emlen S. and Wrege P. 1991. Breeding biology of white-fronted bee-eaters at Nakuru—the influence of helpers on breeder fitness. *Journal of Animal Ecology*, **60**, 309–26.

Engen S. and Saether B.E. 1998. Stochastic population models: some concepts, definitions and results. *Oikos*, **83**, 345–52.

Engen S., Bakke Ø. and Islam A. 1998. Demographic and environmental stochasticity—concepts and definitions. *Biometrics*, **54**, 39–45.

Engen S., Lande R. and Saether B.E. 2003. Demographic stochasticity and Allee effects in populations with two sexes. *Ecology*, **84**, 2378–86.

Estes J., Tinker M., Williams T., and Doak D. 1998. Killer whale predation on sea otters linking oceanic and nearshore ecosystems. *Science*, **282,** 473–6.

Etienne R., Wertheim B., Hemerik L., Schneider P., and Powell J. 2002. The interaction between dispersal, the Allee effect and scramble competition affects population dynamics. *Ecological Modelling*, **148**, 153–68.

Ewing H.E. 1943. Continued fertility in female box turtles following mating. *Copeia*, No. 2, 112–14.

Fagan W., Lewis M., Neubert M., and van den Driessche P. 2002. Invasion theory and biological control. *Ecology Letters*, **5,** 148–57.

Fagan W.F., Lewis M., Neubert M.R., Aumann C., Apple J.L., and Bishop J.G. 2005. When can herbivores slow or reverse the spread of an invading plant? A test case from Mount St. Helens. *American Naturalist*, **166**, 669–85.

Fath G. 1998. Propagation failure of travelling waves in a discrete bistable medium. *Physica D*, **116**, 176–90.

Fausto J.A. Jr, Eckhart V.M., and Geber M.A. 2001. Reproductive assurance and the evolutionary ecology of self-pollination in *Clarkia xantiana* (Onagraceae). *American Journal of Botany*, **88**, 1794–800.

Fauvergue X., Hopper K.R., and Antolin M.F. 1995. Mate finding via a trail sex-pheromone by a parasitoid wasp. *Proceedings of the National Academy of Sciences of the United States of America*, **92**, 900–4.

Fauvergue X., Malausa J.-C., Giuge L., and Courchamp F. 2007. Invading parasitoids suffer no Allee effect: a manipulative field experiment. *Ecology*.

Ferdy J.B. and Molofsky J. 2002. Allee effect, spatial structure and species coexistence. *Journal of Theoretical Biology*, **217**, 413–24.

Ferdy J.B., Austerlitz F., moret J., Gouyon P.H., and Godelle B. 1999. Pollinator-induced density dependence in deceptive species. *Oikos*, **87**, 549–60.

Festa-Bianchet M., Gaillard J.M., and Côte S.D. 2003. Variable age structure and apparent density dependence in survival of adult ungulates. *Journal of Animal Ecology*, **72**, 640–9.

Fiegna F. and Velicer G.J. 2003. Competitive fates of bacterial social parasites: persistence and self-induced extinction of *Myxococcus xanthus* cheaters. *Proceedings of the Royal Society of London B*, **270**, 1527–34.

Fischer J. and Lindenmayer D. 2000. An assessment of the published results of animal reloca-
tions. *Biological Conservation*, **96**, 1–11.

Fischer M. and Matthies D. 1998a. Effects of population size on performance in the rare plant
Gentianella germanica. Journal of Ecology, **86**, 195–204.

Fischer M. and Matthies D. 1998b. RAPD variation in relation to population size and plant
fitness in the rare *Gentianella germanica* (Gentianaceae). *American Journal of Botany*,
85, 811–19.

Fischer M., Hock M., and Paschke M. 2003. Low genetic variation reduces cross-compatibility and offspring fitness in populations of a narrow endemic plant with a self-incompatibility system. *Conservation Genetics*, **4**, 325–36.

Fischer M., Husi R., Prati D., Peintinger M., van Kleunen M., and Schmid B. 2000b. RAPD
variation among and within small and large populations of the rare clonal plant *Ranunculus
reptans* (Ranunculaceae). *American Journal of Botany*, **87**, 1128–37.

Fischer M., van Kleunen M., and Schmid B 2000a. Genetic Allee effects on performance,
plasticity and developmental stability in a clonal plant. *Ecology Letters*, **3**, 530–39.

Folt C. 1987. An experimental analysis of the costs and benefits of zooplankton aggregation.
In W. Kerfoot and A. Sih, eds. *Predation: direct and indirect impacts on aquatic communities*, pp. 300–15. University Press of New England.

Forsyth S. 2003. Density-dependent seed set in the Haleakala silversword: evidence for an
Allee effect. *Oecologia*, **136**, 551–7.

Foster W. and Traherne J. 1981. Evidence for the dilution effect in the selfish herd from fish
predation on a marine insect. *Nature*, **293**, 466–7.

Fowler C. and Baker J. 1991. A review of animal population dynamics at extremely reduced
population levels. *Reports to the International Whaling Commission*, **41**, 545–54.

Fowler M.S. and Ruxton G.D. 2002. Population dynamic consequences of Allee effects.
Journal of Theoretical Biology, **215**, 39–46.

Frank K. and Brickman D. 2001. Comtemporary management issues confronting fisheries
science. *Journal of Sea Research*, **45**, 173–87.

Frank K., Petrie B., Choi J., and Leggett W. 2005. Trophic cascades in a formerly cod-domi-
nated system. *Science*, **308**, 1621–3.

Frank K.T. and Brickman D. 2000. Allee effects and compensatory population dynamics within a stock complex. *Canadian Journal of Fisheries and Aquatic Sciences*, **57**,
513–17.

Frankham R. 1995. Effective population-size adult-population size ratios in wildlife—a
review. *Genetical Research*, **66**, 95–107.

Frankham R. 1996. Relationship of genetic variation to population size in wildlife.
Conservation Biology, **10**, 1500–8.

Frankham R. and Ralls K. 1998. Inbreeding leads to extinction. *Nature*, **392**, 441–2.

Frankham R., Ballou J.D., and Briscoe D.A. 2002. *Introduction to conservation genetics*.
Cambridge University Press, Cambridge.

Franklin I. 1980. Evolutionary change in small populations. In M. Soule and B. Wilcox,
eds. *Conservation biology: an evolutionary-ecological perspective*, pp. 135–49. Sinauer
Associates, Sunderland, Massachusetts.

Freckleton R. 2000. Biological control as a learning process. *Trends in Ecology & Evolution,* **15,** 263–4.

Freckleton R.P., Watkinson A.R., Green R.E. and Sutherland W.J. 2006. Census error and the detection of density dependence. *Journal of Animal Ecology,* **75,** 837–51.

Fretwell S. 1977. Is the dickcissel a threatened species? *American Birds,* **31,** 923–32.

Frusher S. and Hoenig J. 2001. Impact of lobster size on selectivity of traps for southern rock lobster (*Jasus edwardsii*). *Canadian Journal of Fisheries and Aquatic Science,* **58,** 2482–9.

Frusher S. and Hoenig J. 2003. Recent developments in estimating fishing and natural mortality and tag reporting rate of lobsters using multi-year tagging models. *Fisheries Research,* **65,** 379–90.

Fu C., Mohn R., and Fanning L. 2001. Why the Atlantic cod (*Gadus morhua*) stock off eastern Nova Scotia has not recovered. *Canadian Journal of Fisheries and Aquatic Science,* **58,** 1613–23.

Galeuchet D.J., Perret C., and Fischer M. 2005. Performance of *Lychnis flos-cuculi* from fragmented populations under experimental biotic interactions. *Ecology,* **86,** 1002–11.

Gardmark A., Enbert K., Ripa J., Laakso J., and Kaitala V. 2003. The ecology of recovery. *Annales Zoologici Fennici,* **40,** 131–40.

Gardner J. 2004. Winter flocking behaviour of speckled warblers and the Allee effect. *Biological Conservation,* **118,** 195–204.

Garrett K.A. and Bowden R.L. 2002. An Allee effect reduces the invasive potential of *Tilletia indica. Phytopathology,* **92,** 1152–9.

Gascoigne J. and Lipcius R. 2004a. Allee effects driven by predation. *Journal of Applied Ecology,* **41,** 801–10.

Gascoigne J. and Lipcius R. 2004b. Allee effects in marine systems. *Marine Ecology Progress Series,* **269,** 49–59.

Gascoigne J. and Lipcius R.N. 2004c. Conserving populations at low abundance: delayed functional maturity and Allee effects in reproductive behaviour of the queen conch *Strombus gigas. Marine Ecology-Progress Series,* **284,** 185–94.

Gascoigne J. and Lipcius R.N. 2005. Periodic dynamics in a two-stage Allee effect model are driven by tension between stage equilibria. *Theoretical Population Biology,* **68,** 237–41.

Gascoigne J., Beadman H.A., Saurel C., and Kaiser M.J. 2005. Density dependence, spatial scale and patterning in sessile biota. *Oecologia,* **145,** 371–81.

Gerber L.R. and Hilborn R. 2001. Catastrophic events and recovery from low densities in populations of otariids: implications for risk of extinction. *Mammal Review,* **31,** 131–50.

Gerritsen J. and Strickler J.R. 1977. Encounter probabilities and community structure in zooplankton: a mathematical model. *Journal of the Fisheries Research Board of Canada,* **34,** 73–82.

Gilchrist H. 1999. Declining thick-billed murre *Uria lomvia* colonies experience higher gull predation rates: an inter-colony comparison. *Biological Conservation,* **87,** 21–9.

Gilpin M.E. 1972. Enriched predator–prey systems: theoretical stability. *Science,* **177,** 902–4.

Ginzburg L., Slobodkin L., Johnson K., and Bindman A. 1982. Quasiextinction probabilities as a measure of impact on population growth. *Risk Analysis*, **21,** 171–81.

Girman D., Mills M., Geffen E., and Wayne R. 1997. A molecular genetic analysis of social structure, dispersal and interpack relationships of the African wild dog (*Lycaon pictus*). *Behavioural Ecology and Sociobiology*, **40,** 187–98.

Glémin S. 2003. How are deleterious mutations purged? Drift versus nonrandom mating. *Evolution*, **57,** 2678–87.

Godfrey D., Lythgoe J., and Rumball D. 1987. Zebra stripes and tiger stripes: the spatial frequency distribution of the pattern compared to that of the background is significant in display and crypsis. *Biological Journal of the Linnean Society*, **32,** 427–33.

Gopalsamy K. and Ladas G. 1990. On the oscillation and asymptotic behavior of $N(t) = N(t)$ $[a + bN(t - \tau) + cN^2(t - \tau)]$. *Quarterly of Applied Mathematics*, **48,** 433–40.

Gordon D.M. 1995. The development of an ant colony's foraging range. *Animal Behaviour*, **49,** 649–59.

Gordon D.M. and Kulig A.W. 1996. Founding, foraging and fighting: colony size and the spatial distribution of harvester ant nests. *Ecology*, **77,** 2393–409.

Gordon D.M. and Kulig A.W. 1998. The effect of neighbours on the mortality of harvester ant colonies. *Journal of Animal Ecology*, **67,** 141–8.

Gotelli N.J. 1998. *A primer of ecology*. Second edition. Sinauer Associates, Sunderland, MA.

Graham H. 1978. Sterile pink bollworm: field releases for population suppression. *Journal of Economic Entomology*, **71,** 233–5.

Grant P. 2005. The priming of periodical cicada life cycles. *Trends in Ecology & Evolution*, **20,** 169–74.

Green R.E. 1997. The influence of numbers released on the outcome of attempts to introduce exotic bird species to New Zealand. *Journal of Animal Ecology*, **66,** 25–35.

Greene C.M. 2003. Habitat selection reduces extinction of populations subject to Allee effects. *Theoretical Population Biology*, **64,** 1–10.

Greene C.M. and Stamps J.A. 2001. Habitat selection at low population densities. *Ecology*, **82,** 2091–200.

Grevstad F.S. 1999a. Experimental invasions using biological control introductions: the influence of release size on the chance of population establishment. *Biological Invasions*, **1,** 313–23.

Grevstad F.S. 1999b. Factors influencing the chance of population establishment: Implications for release strategies in biocontrol. *Ecological Applications*, **9,** 1439–47.

Griffith B., Scott J., Carpenter J., and Reed C. 1989. Translocation as a species conservation tool—status and strategy. *Science*, **245,** 477–80.

Grimm V. and Railsback S.F. 2005. *Individual-based modeling and ecology*. Princeton University Press, Princeton and Oxford.

Grindeland J., Sletvold N., and Ims R. 2005. Effects of floral display size and plant density on pollinator visitation rate in a natural population of *Digitalis purpurea*. *Functional Ecology*, **19,** 383–90.

Groom M. 1998. Allee effects limit population viability of an annual plant. *American Naturalist*, **151**, 487–96.

Gross J.E., Shipley L.A., Hobbs N.T., Spalinger D.E., and Wunder B.A. 1993. Functional response of herbivores in food-concentrated patches: tests of a mechanistic model. *Ecology*, **74**, 778–91.

Grünbaum D. and Veit R. 2003. Black-browed albatrosses foraging on Antarctic krill: density-dependence through local enhancement? *Ecology*, **84**, 3265–75.

Gruntfest Y., Arditi R., and Dombrovsky Y. 1997. A fragmented population in a varying environment. *Journal of Theoretical Biology*, **185**, 539–47.

Gulati R.D. and van Donk E. 2002. Lakes in the Netherlands, their origin, eutrophication and restoration: state-of-the-art review. *Hydrobiologia*, **478**, 73–106.

Gustafson J. and Crews D. 1981. Effect of group size and physiological state of a cagemate on reproduction in the parthenogenetic lizard *Cnemidophorous uniparens*. *Behavioural Ecology and Sociobiology*, **8**, 267–72.

Gyllenberg M. and Parvinen K. 2001. Necessary and sufficient conditions for evolutionary suicide. *Bulletin of Mathematical Biology*, **63**, 981–93.

Gyllenberg M., Hemminki J., and Tammaru T. 1999. Allee effects can both conserve and create spatial heterogeneity in population densities. *Theoretical Population Biology*, **56**, 231–42.

Gyllenberg M., Osipov A.V., and Söderbacka G. 1996. Bifurcation analysis of a metapopulation model with sources and sinks. *Journal of Nonlinear Science*, **6**, 329–66.

Gyllenberg M., Parvinen K., and Dieckmann U. 2002. Evolutionary suicide and evolution of dispersal in structured metapopulations. *Journal of Mathematical Biology*, **45**, 79–105.

Hackney E.E. and McGraw J.B. 2001. Experimental demonstration of an Allee effect in American ginseng. *Conservation Biology*, **15**, 129–36.

Hadjiavgousti D. and Ichtiaroglou S. 2004. Existence of stable localized structures in population dynamics through the Allee effect. *Chaos, Solitons & Fractals*, **21**, 119–31.

Hadjiavgousti D. and Ichtiaroglou S. 2006. Allee effect in population dynamics: existence of breather-like behavior and control of chaos through dispersal. *International Journal of Bifurcation and Chaos*, **16**, 2001–12.

Haig D. and Westoby M. 1988. On limits to seed production. *American Naturalist*, **131**, 757–9.

Hall S.R., Duffy M.A., and Cáceres C.E. 2005. Selective predation and productivity jointly drive complex behavior in host-parasite systems. *American Naturalist*, **165**, 70–81.

Halliday T. 1980. The extinction of the passenger pigeon *Ectopistes migratorius* and its relevance to contemporary conservation. *Biological Conservation*, **17**, 157–62.

Hamilton J. 1948. Effect of present-day whaling on the stock of whales. *Nature*, **161**, 913–14.

Hamilton W. 1971. Geometry for the selfish herd. *Journal of Theoretical Biology*, **31**, 295–311.

Hanski I. and Gaggiotti O.E. 2004. *Ecology, genetics, and evolution of metapopulations*. Elsevier Academic Press.

Hanski I.A. and Gilpin M.E. 1997. *Metapopulation biology: ecology, genetics, and evolution.* Academic Press, San Diego.

Harcourt R. 1992. Factors affecting early mortality in the South-American fur-seal (*Arctocephalus-australis*) in Peru—density-related effects and predation. *Journal of Zoology,* **226**, 259–70.

Hassell M.P. 2000. *The spatial and temporal dynamics of host-parasitoid interactions.* Oxford University Press, Oxford.

Hassell M.P., Lawton J.H., and Beddington J.R. 1976. The components of arthropod predation. I. The prey death rate. *Journal of Animal Ecology,* **45**, 135–64.

Hastings A. 1996. Models of spatial spread: a synthesis. *Biological Conservation,* **78**, 143–8.

Hauser L., Adcock G.J., Smith P.J., Ramirez J.H.B., and Carvalho G.R. 2002. Loss of microsatellite diversity and low effective population size in an overexploited population of New Zealand snapper (*Pagrus auratus*). *Proceedings of the National Academy of Sciences of the United States of America,* **99**, 11742–7.

Hawkins B.A., Thomas M.B., and Hochberg M.E. 1993. Refuge theory and biological control. *Science,* **262**, 1426–32.

Hays G. 2004. Good news for sea turtles. *Trends in Ecology & Evolution,* **19**, 349–51.

Hearn G., Berghaier R., and George D. 1996. Evidence for social enhancement reproduction two Eulemur species. *Zoo Biology,* **15**, 1–12.

Hedgecock D. 1994. Does variance in reproductive success limit effective population sizes of marine organisms? In AR Beaumont, ed. *Genetics and evolution of aquatic organisms,* pp. 122–34. Chapman and Hall, London.

Hedrick P.W. 2005. Genetics of populations. Third Edition. Jones and Bartlett Publishers, Sudbury, Massachusetts.

Hegazy E. and Khafagi W. 2005. Gregarious development of the solitary endo-parasitoid, *Microplitis rufiventris* in its habitual host, *Spodoptera littoralis. Journal of Applied Entomology,* **129**, 134–41.

Herlihy C.R. and Eckert C.G. 2002. Genetic cost of reproductive assurance in a self-fertilizing plant. *Nature,* **416**, 320–3.

Heschel M.S. and Paige K.N. 1995. Inbreeding depression, environmental stress, and population size variation in scarlet gilia (*Ipomopsis aggregata*). *Conservation Biology,* **9**, 126–33.

Hethcote H.W. 2000. The mathematics of infectious diseases. *SIAM Review,* **42**, 599–653.

Hiddink J., Hutton T., Jennings S., and Kaiser M. 2006a. Predicting the effects of area closuers and fishing effort restrictions on the production, biomass and species richness of benthic invertebrate communities. *ICES Journal of Marine Science,* **63**, 822–30.

Hiddink J.G., Jennings S., Kaiser M.J., Queirós A.M., Duplisea D.E., and Piet G.J. 2006b. Cumulative impacts of seabed trawl disturbance on benthic biomass, production and species richness in different habitats. *Canadian Journal of Fisheries and Aquatic Science,* **63**, 721–36.

Higgins S., Richardson D., and Cowling R. 2000. Using a dynamics landscape model for planning the management of alien plant invasions. *Ecological Applications,* **10**, 1833–48.

Hightower J., Jackson J., and Pollock K. 2001. Use of telemetry methods to estimate natural and fishing mortality of striped bass in Lake Gaston, North Carolina. *Transactions of the American Fisheries Society*, **130**, 557–67.

Hilborn R. and Liermann M. 1998. Standing on the shoulders of giants: learning from experience in fisheries. *Review in Fish Biology and Fisheries*, **8**, 273–83.

Hilborn R. and Walters C. 1992. *Quantitative fisheries stock assessment: choice, dynamics and uncertainty*. Chapman and Hall, New York.

Hilker F.M., Lewis M.A., Seno H., Langlais M., and Malchow H. 2005. Pathogens can slow down or reverse invasion fronts of their hosts. *Biological Invasions*, **7**, 817–32.

Hines A., Jivoff P., Bushmann P., *et al.* 2003. Evidence for sperm limitation in the blue crab, *Callinectes sapidus*. *Bulletin of Marine Science*, **72**, 287–310.

Hissmann K. 1990. Strategies of mate finding in the European field cricket (*Gryllus campestris*) at different population densities: a field study. *Ecological Entomology*, **15**, 281–91.

Hoarau G., Boon E., Jongma D.N., *et al.* 2005. Low effective population size and evidence for inbreeding in an overexploited flatfish, plaice (*Pleuronectes platessa* L.). *Proceedings of the Royal Society B*, **272**, 497–503.

Hobday A.J., Tegner M.J., and Haaker P.L. 2001. Over-exploitation of a broadcast spawning marine invertebrate: Decline of the white abalone. *Reviews in Fish Biology and Fisheries*, **10**, 493–514.

Hoddle M.S. 1991. Lifetable construction for the gorse seed weevil, *Apion ulicis* (Forster) (Coleoptera, Apionidae) before gorse pod dehiscence, and life-history strategies of the weevil. *New Zealand Journal of Zoology*, **18**, 399–404.

Hoeksema J.D. and Bruna E.M. 2000. Pursuing the big questions about interspecific mutualism: a review of theoretical approaches. *Oecologia*, **125**, 321–30.

Hoglund J. 1996. Can mating systems affect local extinction risks? Two examples of lek-breeding waders. *Oikos*, **77**, 184–8.

Hölldobler B. and Wilson E.O. 1990. *The ants*. Springer Verlag, Heidelberg-Berlin.

Holling C.S. 1959. Some characteristics of simple types of predation and parasitism. *Canadian Entomologist*, **91**, 385–98.

Hopf F.A. and Hopf F.W. 1985. The role of the Allee effect in species packing. *Theoretical Population Biology*, **27**, 27–50.

Hoppensteadt F.C. 1982. *Mathematical methods of population biology*. Cambridge University Press, Cambridge.

Hopper K.R. and Roush R.T. 1993. Mate finding, dispersal, number released, and the success of biological-control introductions. *Ecological Entomology*, **18**, 321–31.

Hubbell S.P. 2001. *The unified neutral theory of biodiversity and biogeography*. Princeton University Press, Princeton.

Hughes T.P. 1994. Catastrophes, phase-shifts, and large-scale degradation of a Caribbean coral-reef. *Science*, **265**, 1547–51.

Hughes W.O.H., Petersen K.S., Ugelvig L.V., *et al.* 2004. Density-dependence and within-host competition in a semelparous parasite of leaf-cutting ants. *BMC Evolutionary Biology*, **4**, Art. No. 45.

Hui C. and Li Z.Z. 2004. Distribution patterns of metapopulation determined by Allee effects. *Population Ecology*, **46**, 55–63.

Huntsman G., Potts J., Mays R., and Vaughan D. 1999. Groupers (Serranidae, Epinephelinae): endangered apex predators of reef communities. *American Fisheries Society Symposium*, **23**, 217–31.

Husband B.C. and Schemske D.W. 1996. Evolution of the magnitude and timing of inbreeding depression in plants. *Evolution*, **50**, 54–70.

Huston M., De Angelis D., and Post W. 1988. New computer models unify ecological theory. *BioScience*, **38**, 682–91.

Hutchings J. 2000. Collapse and recovery of marine fishes. *Nature*, **406**, 882–5.

Hutchings J. 2001. Influence of population decline, fishing, and spawner variability on the recovery of marine fishes. *Journal of Fish Biology*, **59** (Supplement A), 303–22.

Hutchings J. and Reynolds J. 2004. Marine fish population collapses: consequences for recovery and extinction risk. *BioScience*, **54**, 297–309.

Hutchinson W.F., van Oosterhout C., Rogers S.I., and Carvalho G.R. 2003. Temporal analysis of archived samples indicates marked genetic change in declining North Sea cod (*Gadus morhua*). *Proceedings of the Royal Society B*, **270**, 2125–32.

Irfanullah H.M. and Moss B. 2005. A filamentous green algae-dominated temperate shallow lake: variations on the theme of clearwater stable states? *Archiv Fur Hydrobiologie*, **163**, 25–47.

Ivlev V.S. 1961. *Experimental ecology of the feeding of fishes*. Yale University Press, New Haven.

Iwahashi O. 1977. Eradication of the melon fly, *Dacus cucurbitae*, from Kume Island, Okinawa, with the sterile insect release method. *Researches on Population Ecology*, **19**, 87–98.

Jabbour H.N., Veldhuizen F.A., Mulley R.C., and Asher G.W. 1994. Effect of exogenous gonadotropins on estrus, the LH surge and the timing and rate of ovulation in red deer (*Cervus-elaphus*). *Journal or Reproduction and Fertility*, **100**, 533–9.

Jackson A.P. 2004. Cophylogeny of the Ficus microcosm. *Biological Reviews*, **79**, 751–68.

Jackson J., Kirby M., Berger W., *et al.* 2001. Historical overfishing and the recent collapse of coastal ecosystems. *Science*, **293**, 629–38.

Jacobs J. 1984. Cooperation, optimal density and low density thresholds: yet another modification of the logistic model. *Oecologia*, **64**, 389–95.

Jang S.R.J. 2006. Allee effects in a discrete-time host-parasitoid model. *Journal of Difference Equations and Applications*, **12**, 165–81.

Janzen D.H. 1971. Seed predation by animals. *Annual Review of Ecology and Systematics*, **2**, 465–92.

Jarvis J., Bennett N., and Spinks A. 1998. Food availability and foraging by wild colonies of Damaraland mole-rats (*Cryptomys damarensis*): implications for sociality. *Oecologia*, **113**, 290–8.

Jennings S. and Kaiser M. 1998. The effects of fishing on marine ecosystems. *Advances in Marine Biology*, **34**, 201–352.

Jonhson D.M., Liebhold A.M., Tobin P.C., and Bjørnstad O.N. 2006. Allee effects and pulsed invasion by the gypsy moth. *Nature*, **444**, 361–3.

Jonsson M., Kindvall O., Jonsell M., and Nordlander G. 2003. Modelling mating success of saproxylic beetles in relation to search behaviour, population density and substrate abundance. *Animal Behaviour*, **65**, 1069–76.

Jordal B.H., Beaver R.A., and Kirkendall L.R. 2001. Breaking taboos in the tropics: incest promotes colonization by wood-boring beetles. *Global Ecology and Biogeography*, **10**, 345–57.

Jutagate T., De Silva S., and Mattson N. 2003. Yield, growth and mortality rate of the Thai river sprat, *Clupeichthys aesarnensis*, in Sirinthorn Reservoir, Thailand. *Fisheries Management and Ecology*, **10,** 221–31.

Kaiser M.J. 2005. Are marine protected areas a red herring or fisheries panacea? *Canadian Journal of Fisheries and Aquatic Science*, **62**, 1194–99.

Kamukuru A., Hecht T., and Mgaya Y. 2005. Effects of exploitation on age, growth and mortality of the blackspot snapper, *Lutjanus fulviflamma*, at Mafia Island, Tanzania. *Fisheries Management and Ecology*, **12,** 45–55.

Kareiva P., Laurance W., and Stapley L. 2002. Irresponsible harvest puts white abalone on verge of extinction. *Trends in Ecology & Evolution*, **17,** 550.

Kaspari M. and Byrne M.M. 1995. Caste allocation in litter *Pheidole*: lessons from plant defense theory. *Behavioral Ecology and Sociobiology*, **37**, 255–63.

Keeling M.J., Jiggins F.M., and Read J.M. 2003. The invasion and coexistence of competing *Wolbachia* strains. *Heredity*, **91**, 382–8.

Keitt T.H., Lewis M.A., and Holt R.D. 2001. Allee effects, invasion pinning, and species' borders. *American Naturalist*, **157**, 203–16.

Keller L.F. and Waller D.M. 2002. Inbreeding effects in wild populations. *Trends in Ecology & Evolution*, **17**, 230–41.

Kent A., Doncaster C.P., and Sluckin T. 2003. Consequences for predators of rescue and Allee effects on prey. *Ecological Modelling*, **162**, 233–45.

Kenward R.E. 1978. Hawks and doves: factors affecting success and selection in goshawk attacks on woodpigeons. *Journal of Animal Ecology*, **47**, 449–60.

Kermack W.O. and McKendrick A.G. 1927. A contribution to the mathematical theory of epidemics. *Proceedings of the Royal Society of London. Series A, Containing Papers of a Mathematical and Physical Character*, **115**, 700–21.

Kéry M., Matthies D., and Spillmann H.H, 2000. Reduced fecundity and offspring performance in small populations of the declining grassland plants *Primula veris* and *Gentiana lutea*. *Journal of Ecology*, **88**, 17–30.

Kiesecker J.M., Blaustein A.R., and Belden L.K. 2001. Complex causes of amphibian population declines. *Nature*, **410,** 681–4.

Kindvall O., Vessby K., Berggren Å., and Hartman G. 1998. Individual mobility prevents an Allee effect in sparse populations of the bush cricket *Metrioptera roeseli*: an experimental study. *Oikos*, **81**, 449–57.

Kiørboe T. 2006. Sex, sex-ratios, and the dynamics of pelagic copepod populations. *Oecologia*, **148**, 40–50.

Klausmeier C. 1999. Regular and irregular patterns in semiarid vegetation. *Science*, **284**, 1826–8.

Klomp H., van Montfort M.A.J., and Tammes P.M.L. 1964. Sexual reproduction and under-population. *Archives Neerlandaises De Zoologie*, Tome XVI, 1, 105–10.

Knapp E., Goedde M., and Rice K. 2001. Pollen-limited reproduction in blue oak: implications for wind pollination in fragmented populations. *Oecologia*, **128**, 48–55.

Knowlton N. 1992. Thresholds and multiple stable states in coral-reef community dynamics. *American Zoologist*, **32**, 674–82.

Koenig C., Coleman F., Collins L., Sadovy Y., and Colin P. 1996. Reproduction in gag (*Mycteroperca microlepis*) in the eastern Gulf of Mexico and the consequences of fishing spawning aggregations. In F. Arreguin-Sanchez, J. Munro, M. Balgos and D. Pauly, eds. *Biology, Fisheries and Culture of Tropical Groupers and Snappers*, pp. 307–23. ICLARM Conference Proceedings.

Koenig W. and Ashley M. 2003. Is pollen limited? The answer is blowin' in the wind. *Trends in Ecology & Evolution*, **18**, 157–9.

Koenig W. and Knops J. 1998. Scale of mast-seeding and tree-ring growth. *Nature*, **396**, 225–6.

Koenig W. and Knops J. 2005. The mystery of masting in trees. *American Scientist*, **93**, 340–7.

Kokko H. and Sutherland W.J. 2001. Ecological traps in changing environments: ecological and evolutionary consequences of a behaviourally mediated Allee effect. *Evolutionary Ecology Research*, **3**, 537–51.

Kolb A. and Lindhorst S. 2006. Forest fragmentation and plant reproductive success: a case study in four perennial herbs. *Plant Ecology*, **185**, 209–20.

Komers P. and Curman G. 2000. The effect of demographic characteristics on the success of ungulate re-introductions. *Biological Conservation*, **93**, 187–93.

Kon H., Noda T., Terazawa K., Koyama H., and Yasaka M. 2005. Evolutionary advantages of mast seeding in *Fagus crenata*. *Journal of Ecology*, **93**, 1148–55.

Kot M. 2001. *Elements of mathematical ecology*. Cambridge University Press, Cambridge.

Kot M., Lewis M.A., and van den Driessche P. 1996. Dispersal data and the spread of invading organisms. *Ecology*, **77**, 2027–42.

Krafsur E. 1998. Sterile insect technique for suppressing and eradicating insect populations: 55 years and counting. *Journal of Agricultural Entomology*, **15**, 303–17.

Krebs J., MacRoberts M., and Cullen J. 1972. Flocking and feeding in the great tit *Parus major*—an experimental study. *Ibis*, **114**, 507–34.

Kulenovic M.R.S. and Yakubu A.A. 2004. Compensatory versus overcompensatory dynamics in density-dependent Leslie models. *Journal of Difference Equations and Applications*, **10**, 1251–65.

Kunin W. 1993. Sex and the single mustard: population density and pollinator effects on seed-set. *Ecology*, **74**, 2145–60.

Kuno E. 1978. Simple mathematical models to describe the rate of mating in insect populations. *Researches on Population Ecology*, **20**, 50–60.

Kuspa A., Plamann L., and Kaiser D. 1992. A-signalling and the cell density requirement for *Myxococcus xanthus* development. *Journal of Bacteriology*, **174**, 7360–9.

Kuussaari M., Saccheri I., Camara M., and Hanski I. 1998. Allee effect and population dynamics in the Glanville fritillary butterfly. *Oikos*, **82**, 384–92.

Lack D. 1968. *Ecological Adaptations for Breeding in Birds*. Methuen and Co. Ltd., London.

Laikre L. and Ryman N. 1991. Inbreeding depression in a captive wolf (*Canis lupus*) population. *Conservation Biology*, **5**, 33–40.

Laiolo P. and Tella J.L. 2007. Erosion of animal cultures in fragmented landscapes. *Frontiers in Ecology and the Environment*, **5**, 68–72.

Lamberson R.H., McKelvey R., Noon B.R., and Voss C. 1992. A dynamic analysis of northern spotted owl viability in a fragmented forest landscape. *Conservation Biology*, **6**, 502–12.

Lamberson R.H., Noon B.R., Voss C., and McKelvey K.S. 1994. Reserve design for territorial species—the effects of patch size and spacing on the viability of the Northern Spotted Owl. *Conservation Biology*, **8**, 185–95.

Lamont B.B., Klinkhamer P.G.L., and Witkowski E.T.F. 1993. Population fragmentation may reduce fertility to zero in *Banksia goodii*—a demonstration of the Allee effect. *Oecologia*, **94**, 446–50.

Lande R. 1988. Genetics and demography in biological conservation. *Science*, **241**, 1455–60.

Lande R. 1993. Risks of population extinction from demographic and environmental stochasticity and random catastrophes. *American Naturalist*, **142**, 911–27.

Lande R. 1998a. Anthropogenic, ecological and genetic factors in extinction and conservation. *Researches on Population Ecology*, **40**, 259–69.

Lande R. 1998b. Demographic stochasticity and Allee effect on a scale with isotropic noise. *Oikos*, **83**, 353–8.

Lande R., Engen S. and Saether B.E. 1994. Optimal harvesting, economic discounting and extinction risk in fluctuating populations. *Nature*, **372**, 88–90.

Lande R., Engen S., and Saether B.E. 1998. Extinction times in finite metapopulation models with stochastic local dynamics. *Oikos*, **83**, 383–9.

Langeland A. and Pedersen T. 2000. A 27-year study of brown trout population dynamics and exploitation in Lake Songsjoen, central Norway. *Journal of Fish Biology*, **57**, 1227–44.

Le Cadre S., Stoeckel S., Bessa-Gomes C., and Machon. In review. N. O brother, where art thou? Tales of intraspecific interactions and resulting Allee effects in plants. *Trends in Ecology & Evolution*.

Lee H. and Hsu C. 2003. Population biology of the swimming crab *Portunus sanguinolentus* in the waters off northern Taiwan. *Journal of Crustacean Biology*, **23**, 691–9.

Lehmann L., Perrin N., and Rousset F. 2006. Population demography and the evolution of helping behaviors. *Evolution*, **60**, 1137–51.

Lennartsson T. 2002. Extinction thresholds and disrupted plant-pollinator interactions in fragmented plant populations. *Ecology*, **83**, 3060–72.

Les D.H., Reinartz J.A., and Esselman E.J. 1991. Genetic consequences of rarity in *Aster furcatus* (Asteraceae), a threatened, self-incompatible plant. *Evolution*, **45**, 1641–50.

Lessios H. 2005. *Diadema antillarum* populations in Panama twenty years following mass mortality. *Coral Reefs*, **24**, 125–7.

Letcher B.H., Priddy J.A., Walters J.R., and Crowder L.B. 1998. An individual-based, spatially-explicit simulation model of the population dynamics of the endangered red-cockaded woodpecker, *Picoides borealis*. *Biological Conservation*, **86**, 1–14.

Leung B., Drake J., and Lodge D. 2004. Predicting invasions: propagule pressure and the gravity of Allee effects. *Ecology*, **85**, 1651–60.

Levitan D.R. and McGovern T.M. 2005. The Allee effect in the sea. In E.A. Norse and L.B. Crowder, eds. *Marine conservation biology: the science of maintaining the sea's biodiversity*, pp. 47–57. Island Press.

Levitan D.R. 1991. Influence of body size and population density on fertilization success and reproductive output in a free-spawning invertebrate. *Biological Bulletin*, **181**, 261–8.

Levitan D.R. 1998. Sperm limitation, sperm competition and sexual selection in external fertilizers. In T. Birkhead and A. Moller, eds. *Sperm competition and sexual selection*, pp. 173–215. Academic Press, San Diego, California.

Levitan D.R. (2002a). Density-dependent selection on gamete traits in three congeneric sea urchins. *Ecology*, **83**, 464–79.

Levitan D.R. (2002b). The relationship between conspecific fertilization success and reproductive isolation among three congeneric sea urchins. *Evolution*, **56**, 1599–1609.

Levitan D.R. and Young C.M. 1995. Reproductive success in large populations—empirical measures and theoretical predictions of fertilization in the sea biscuit *Clypeaster rosaceus*. *Journal of Experimental Marine Biology and Ecology*, **190**, 221–41.

Levitan D.R., Sewell M.A., and Chia F.S. 1992. How distribution and abundance influence fertilization success in the sea urchin *Strongylocentrotus franciscanus*. *Ecology*, **73**, 248–54.

Lewis M.A. and Kareiva P. 1993. Allee dynamics and the spread of invading organisms. *Theoretical Population Biology*, **43**, 141–58.

Lewis M.A. and van den Driessche P. 1993. Waves of extinction from sterile insect release. *Mathematical Biosciences*, **116**, 221–47.

Liebhold A. and Bascompte J. 2003. The Allee effect, stochastic dynamics and the eradication of alien species. *Ecology Letters*, **6**, 133–40.

Liermann M. and Hilborn R. 1997. Depensation in fish stocks: a hierarchic Bayesian meta-analysis. *Canadian Journal of Fisheries and Aquatic Sciences*, **54**, 1976–84.

Liermann M. and Hilborn R. 2001. Depensation: evidence, models and implications. *Fish and Fisheries*, **2**, 33–58.

Ligon J.D. and Burt D.B. 2004. Evolutionary origins. In W.D. Koenig and J.L. Dickinson, eds. *Ecology and evolution of cooperative breeding in birds*, pp. 5–34. Cambridge University Press, Cambridge.

Lloyd D.G. 1992. Self- and cross-fertilization in plants. II. The selection of self-fertilization. *International Journal of Plant Sciences*, **153**, 370–80.

Long J.L. 1981. *Introduced birds of the world*. Universe, New York.

Ludwig D. 1998. Management of stocks that may collapse. *Oikos*, **83**, 397–402.

Luijten S.H., Dierick A., Gerard J., Oostermeijer B., Raijmann L.E.L., and Den Nijs H.C.M. 2000. Population size, genetic variation, and reproductive success in a rapidly declining, self-incompatible perennial (*Arnica montana*) in the Netherlands. *Conservation Biology*, **14**, 1776–87.

Lynch M., Conery J., and Burger R. 1995. Mutation accumulation and the extinction of small populations. *American Naturalist*, **146**, 489–518.

MacArthur R. and Wilson E. 1967. *The Theory of Island Biogeography*. Princeton University Press.

MacDiarmid A.B. and Butler M.J. 1999. Sperm economy and limitation in spiny lobsters. *Behavioural Ecology and Sociobiology*, **46**, 14–24.

MacGowan K. and Woolfenden G. 1989. A sentinel system in the Florida scrub jay. *Animal Behaviour*, **37**, 1000–6.

MacMahon J.A., Mull J.F., and Crist T.O. 2000. Harvester ants (*Pogonomyrmex* spp.): community and ecosystem influences. *Annual Review of Ecology and Systematics*, **31**, 265–91.

Madsen T., Stille B., and Shine R., 1996. Inbreeding depression in an isolated population of adders *Vipera berus*. *Biological Conservation*, **75**, 113–18.

Maeto K. and Ozaki K. 2003. Prolonged diapause of specialist seed-feeders makes predator satiation unstable in masting of *Quercus crispula*. *Oecologia*, **137**, 392–8.

Magrath R. 2001. Group breeding dramatically increases reproductive success of yearling but not older female scrubwrens: a model for cooperatively breeding birds? *Journal of Animal Ecology*, **70**, 370–85.

Machado C.A., Robbins N., Gilbert M.T.P., and Herre E.A. 2005. Critical review of host specificity and its coevolutionary implications in the fig/fig-wasp mutualism. *Proceedings of the National Academy of Sciences of the United States of America*, **102**, 6558–65.

Maia A.C.D. and Schlindwein C. 2006. *Caladium bicolor* (Araceae) and *Cyclocephata celata* (Coleoptera, Dynastinae): a well-established pollination system in the northern Atlantic Rainforest of Pernambuco, Brazil. *Plant Biology*, **8**, 529–34.

Marshall L.M. 1992. Survival of juvenile queen conch, *Strombus gigas*, in natural habitats: impact of prey, predator and habitat features. PhD thesis, College of William and Mary.

Martínez J.A. and Zuberogoitia I. 2001. The response of the Eagle Owl (*Bubo bubo*) to an outbreak of the rabbit haemorrhagic disease. *Journal für Ornithologie*, **142**, 204–11.

Martinez-Munoz M. and Ortega-Salas A. 2001. Growth and mortality of the bigmouth sole, *Hippoglossina stomata*, off the western coast of Baja California, Mexico. *Bulletin of Marine Science*, **69**, 1109–19.

May R.M. 1974. Biological populations with nonoverlapping generations: stable points, stable cycles and chaos. *Science*, **186**, 645–7.

May R.M. 1977a. Togetherness among Schistosomes: its effects on the dynamics of the infection. *Mathematical Biosciences*, **35**, 301–43.

May R.M. 1977b. Thresholds and breakpoints in ecosystems with a multiplicity of stable states. *Nature*, **269**, 471–7.

May R.M. and Anderson R.M. 1979. Population biology of infectious diseases: part I. *Nature*, **280**, 361–7.

McCallum H., Barlow N., and Hone J. 2001. How should pathogen transmission be modelled? *Trends in Ecology & Evolution*, **16**, 295–300.

McCarthy M.A. 1997. The Allee effect, finding mates and theoretical models. *Ecological Modelling*, **103**, 99–102.

McCauley E., Kendall B.E., Janssen A., *et al.* 2000. Inferring colonisation processes from population dynamics in spatially structured predator–prey systems. *Ecology*, **81**, 3350–61.

McComb K., Moss C., Durant S., Baker L., and Saylalel S. 2001. Matriachs as repositories of social knowledge in African elephants. *Science*, **292**, 491–4.

McJunkin J.W., Zelmer D.A., and Applegate R.D. 2005. Population dynamics of wild turkeys in Kansas (*Meleagris gallopavo*): theoretical considerations and implications of rural mail carrier survey (RMCS) data. *American Midland Naturalist*, **154**, 178–87.

McKelvey R., Noon B., and Lamberson R., 1993. Conservation planning for species occupying fragmented landscapes: the case of the northern spotted owl. In P. Kareiva, J. Kingsolver, and R. Huey, eds. *Biotic interactions and global change*, pp. 424–50. Sinauer Associates Inc., Sunderland, Massachusetts.

Mehanna S. and El Ganainy A. 2003. Population dynamics of *Sardinella jussieui* (Valenciennes, 1847) in the Gulf of Suez, Egypt. *Indian Journal of Fisheries*, **50**, 67–71.

Menges E.S. 1991. Seed germination percentage increases with population size in a fragmented prairie species. *Conservation Biology*, **5**, 158–64.

Miller T. 2001. Control of pink bollworm. *Pesticide Outlook*, **April 2001,** 68–70.

Mills L.S. and Allendorf F.W. 1996. The one-migrant-per-generation rule in conservation and management. *Conservation Biology*, **10**, 1509–18.

Mills L.S. and Smouse P.E. 1994. Demographic consequences of inbreeding in remnant populations. *American Naturalist*, **144**, 412–31.

Milner-Gulland E.J., Bukreeva O.M., Coulson T., *et al.* 2003. Reproductive collapse in saiga antelope harems. *Nature*, **422**, 135.

Møller A.P. 1987. Advantages and disadvantages of coloniality in the swallow *Hirando rustica*. *Animal Behaviour*, **35**, 819–32.

Møller A.P. and Legendre S. 2001. Allee effect, sexual selection and demographic stochasticity. *Oikos*, **92**, 27–34.

Møller A.P. and Thornhill R. 1998. Male parental care, differential parental investment by females and sexual selection. *Animal Behaviour*, **55**, 1507–15.

Moeller D.A. and Geber M.A. 2005. Ecological context of the evolution of self-pollination in *Clarkia xantiana*: population size, plant communities, and reproductive assurance. *Evolution*, **59**, 786–99.

Moore D.M. and Lewis H. 1965. The evolution of self-pollination in *Clarkia xantiana*. *Evolution*, **19**, 104–14.

Mooring M., Fitzpatrick T., Nishihira T., and Reisig D. 2004. Vigilance, predation risk and the Allee effect in desert bighorn sheep. *Journal of Wildlife Management*, **68**, 519–32.

Morales-Bojórquez E. and Nevárez-Martínez M.O. 2005. Spawner-recruit patterns and investigation of Allee effect in pacific sardine (*Sardinops sagax*) in the Gulf of California, Mexico. *California Cooperative Oceanic Fisheries Investigations Reports*, **46**, 161–74.

Morgan L., Botsford L., Wing S., and Smith B. 2000. Spatial variability in growth and mortality of the red sea urchin, *Strongylocentrotus franciscanus*, in northern California. *Canadian Journal of Fisheries and Aquatic Science*, **57**, 980–92.

Morgan R.A., Brown J.S., and Thorson J.M., 1997. The effect of spatial scale on the functional response of fox squirrels. *Ecology*, **78**, 1087–97.

Morgan M.T., Wilson W.G., and Knight T.M., 2005. Plant population dynamics, pollinator foraging, and the selection of self-fertilization. *American Naturalist*, **166**, 169–83.

Morozov A., Petrovskii S., and Li B.L. 2004. Bifurcations and chaos in a predator–prey system with the Allee effect. *Proceedings of the Royal Society of London Series B-Biological Sciences*, **271**, 1407–14.

Morris D. 2002. Measuring the Allee effect: positive density dependence in small mammals. *Ecology*, **83**, 14–20.

Morton T., Haefner J., Nugala V., Decino R., and Mendes L. 1994. The selfish herd revisited: do simple movement rules reduce relative predation risk? *Journal of Theoretical Biology*, **167**, 37–79.

Mosimann J.E. 1958. The evolutionary significance of rare matings in animal populations. *Evolution*, **12**, 246–61.

Muchhala N. 2006. Nectar bat stows huge tongue in its rib cage. *Nature*, **444**, 701–2.

Muir W.M. and Howard R.D. 1999. Possible ecological risks of transgenic organism release when transgenes affect mating success: sexual selection and the Trojan gene hypothesis. *Proceedings of the National Academy of Sciences of the United States of America*, **96**, 13853–6.

Mumme R. and DeQueiroz A. 1985. Individual contributions to cooperative behavior in the acorn woodpecker—effects of reproductive status, sex, and group-size. *Behaviour*, **95**, 290–313.

Murdoch W.W. and Oaten A. 1975. Predation and population stability. *Advances in Ecological Research*, **9**, 1–131.

Murdoch W.W., Kendall B.E., Nisbet R.M., Briggs C.J., McCauley E., and Bolser R. 2002. Single-species models for many-species food webs. *Nature*, **417**, 541–3.

Murray J.D. 1993 *Mathematical biology*. Second edition. Springer-Verlag, Berlin and Heidelberg.

Myers R. and Barrowman N. 1996. Is fish recruitment related to spawner abundance? *Fishery Bulletin*, **94**, 707–24.

Myers R., Barrowman N., Hutchings J., and Rosenberg A. 1995. Population dynamics of exploited fish stocks at low population levels. *Science*, **269**, 1106–8.

Myers R., Hutchings J., and Barrowman N. 1997. Why do fish stocks collapse? The example of cod in Atlantic Canada. *Ecological Applications*, **7**, 91–106.

Myers R.A. and Worm B. 2003. Rapid worldwide depletion of predatory fish communities. *Nature*, **423**, 280–3.

Myers R.A., Barrowman N.J., Hutchings J.A., and Rosenberg A.A. 1995. Population-dynamics of exploited fish stocks at low population-levels. *Science*, **269**, 1106–8.

Newman D. and Pilson D. 1997. Increased probability of extinction due to decreased genetic effective population size: experimental populations of *Clarkia pulchella*. *Evolution*, **51**, 354–62.

Newsome A. and Noble I. 1986. Ecological and physiological characters of invading species. In R. Groves and J. Burdon, eds. *Ecology of biological invasions*, pp. 1–20. Cambridge University Press.

Nilsson S. and Wastljung U. 1987. Seed predation and cross-pollination in mast-seeding beech (*Fagus sylvatica*) patches. *Ecology*, **68**, 260–5.

Nunney L. 1993. The influence of mating system and overlapping generations on effective population size. *Evolution*, **47**, 1329–41.

Odum E.P. 1953. *Fundamentals of Ecology*. Saunders, Philadelphia, Pennsylvania.

Odum H.T. and Allee W.C. 1954. A note on the stable point of populations showing both intraspecific cooperation and disoperation. *Ecology*, **35**, 95–7.

Ofori-Danson P., de Graaf G., and Vanderpuye C. 2002. Population parameter estimates for *Chrysichthys auratus* and *C. nigrodigitatus* (Pisces: Claroteidae) in Lake Volta, Ghana. *Fisheries Research*, **54**, 267–77.

Olaya-Nieto C. and Appeldoorn R. 2004. Age and growth of striped mojarra, *Eugerres plumieri* (Cuvier), in the Cienaga Grande De Santa Marta, Colombia. *Proceedings of the Gulf and Caribbean Fisheries Institute*, **55**, 337–47.

Olsen E.M., Heino M., Lilly G.R., *et al.* 2004. Maturation trends indicative of rapid evolution preceded the collapse of northern cod. *Nature*, **428**, 932–5.

Oostermeijer J., Luijten S., and den Nijs J. 2003. Integrating demographic and genetic approaches in plant conservation. *Biological Conservation*, **113**, 389–98.

Oostermeijer J.G.B. 2000. Population viability analysis of the rare *Gentiana pneumonanthe*: importance of genetics, demography, and reproductive biology. In A.G. Young and G.M. Clarke, eds. *Genetics, demography and viability of fragmented populations*, pp. 313–34. Cambridge University Press, Cambridge.

Oostermeijer J.G.B., Luijten S.H., and den Nijs J.C.M. 2003. Integrating demographic and genetic approaches in plant conservation. *Biological Conservation*, **113**, 389–98.

Oro D. and Ruxton G. 2001. The formation and growth of seabird colonies: Audouin's gull as a case study. *Journal of Animal Ecology*, **70**, 527–35.

Oro D., Martinez-Abrain A., Paracuellos M., Nevado J., and Genovart M., 2006. Influence of density dependence on predator–prey seabird interactions at large spatio-temporal scales. *Proceedings of the Royal Society of London B*, **273**, 379–83.

Ouborg N.J., van Treunen R., and van Damme J.M.M. 1991. The significance of genetic erosion in the process of extinction II. Morphological variation and fitness components in populations of varying size of *Salvia pratensis* L. and *Scabiosa columbaria* L. *Oecologia*, **86**, 359–67.

Owen M.R. and Lewis M.A. 2001. How predation can slow, stop or reverse a prey invasion. *Bulletin of Mathematical Biology*, **63**, 655–84.

Padrón V. and Trevisan M.C. 2000. Effect of aggregating behavior on population recovery on a set of habitat islands. *Mathematical Biosciences*, **165**, 63–78.

Parker D., Haaker P., and Togstad H. 1992. Case histories for three species of California abalone, *Haliotis corrugata*, *H. fulgens* and *H. cracherodii*. In S. Shepherd, M. Tegner and G.D. Proo, eds. *Abalone of the world: biology, fisheries and culture*, pp. 384–94. Fishing News Books, Oxford.

Parvinen K. 2005. Evolutionary suicide. *Acta Biotheoretica*, **53**, 241–64.

Pedersen B., Hanslin H.M., and Bakken S. 2001. Testing for positive density-dependent performance in four bryophyte species. *Ecology*, **82**, 70–88.

Peterman R. 1980. Dynamics of native Indian food fisheries on salmon in British Columbia. *Canadian Journal of Fisheries and Aquatic Science*, **37**, 561–6.

Peterson C. and Levitan D.R. 2001. The Allee effect: a barrier to recovery by exploited species. In J.D. Reynolds, G.M. Mace, K.H. Redford, and J.G. Robinson, eds. *Conservation of exploited species*, pp. 281–300. Cambridge University Press, Cambridge.

Petrovskii S.H., Malchow H., and Li B.L. 2005. An exact solution of a diffusive predator–prey system. *Proceedings of the Royal Society A—Mathematical, Physical and Engineering Sciences*, **461**, 1029–53.

Petrovskii S.V. and Li B.L. 2003. An exactly solvable model of population dynamics with density-dependent migrations and the Allee effect. *Mathematical Biosciences*, **186**, 79–91.

Petrovskii S.V., Malchow H., Hilker F.M., and Venturino E. 2005b. Patterns of patchy spread in deterministic and stochastic models of biological invasion and biological control. *Biological Invasions*, **7**, 771–93.

Petrovskii S.V., Morozov A.Y., and Li B.L. 2005a. Regimes of biological invasion in a predator–prey system with the Allee effect. *Bulletin of Mathematical Biology*, **67**, 637–61.

Petrovskii S.V., Morozov A.Y., and Venturino E. 2002. Allee effect makes possible patchy invasion in a predator–prey system. *Ecology Letters*, **5**, 345–52.

Pfister C.A. and Bradbury A. 1996. Harvesting red sea urchins: recent effects and future predictions. *Ecological Applications*, **6**, 298–310.

Philip J.R. 1957. Sociality and sparse populations. *Ecology*, **38**, 107–11.

Phillips B.F., Booth J.D., Cobb S.J., Jeffs A., and McWilliam P. 2006. Larval and post-larval ecology. In B.F. Phillips, ed. *Lobsters: biology, management, aquaculture and fisheries*, pp. 231–62. Blackwell Publishing, Oxford.

Pierce S., Ceriani R., Villa M., and Cerabolini B. 2006. Quantifying relative extinction risks and targeting intervention for the orchid flora of a natural park in the European prealps. *Conservation Biology*, **20**, 1804–10.

Pichon G., Awono-Ambene H., and Robert V. 2000. High heterogeneity in the number of *Plasmodium falciparum* gametocytes in the bloodmeal of mosquitos fed on the same host. *Parasitology*, **121**, 115–20.

Post E., Levin S.A., Iwasa Y., and Stenseth N.C. 2001. Reproductive asynchrony increases with environmental disturbance. *Evolution*, **55**, 830–4.

Post J., Sullivan M., Cox S., Lester N., Walters C., Parkinson E., Paul A., Jackson L., and Shuter B. 2002. Canada's recreational fisheries: the invisible collapse? *Fisheries*, **27**, 6–17.

Prince J. 2005. Combating the tyranny of scale for haliotids: micro-management for microstocks. *Bulletin of Marine Science*, **76**, 557–78.

Prober S.M. and Brown A.H.D. 1994. Conservation of the grassy white box woodlands: population genetics and fragmentation of *Eucalyptus albens*. *Conservation Biology*, **8**, 1003–13.

Proverbs M., Newton J., and Logan D. 1977. Codling moth control by the sterility method in twenty-one British Columbia orchards. *Journal of Economic Entomology*, **70**, 667–71.

Punt A. and Smith A. 2001. The gospel of maximum sustainable yield in fisheries management: birth, crucifixion and reincarnation. In J. Reynolds, G. Mace, K. Redford, and J. Robinson, eds. *Conservation of Exploited Species*, pp. 41–66. Cambridge University Press, Cambridge.

Quenette P.-Y. 1990. Functions of vigilance behaviour in mammals: a review. *Acta Oecologica*, **11**, 801–18.

Quinn J.F., Wing S.R., and Botsford L.W. 1993. Harvest refugia in marine invertebrate fisheries—models and applications to the red-sea urchin, *Strongylocentrotus-franciscanus*. *American Zoologist*, **33**, 537–50.

Ralls K., Brugger K., and Ballou J. 1979. Inbreeding and juvenile mortality in small populations of ungulates. *Science*, **206**, 1101–3.

Rangeley R. and Kramer D. 1998. Density-dependent antipredator tactics and habitat selection in juvenile pollock. *Ecology*, **79**, 943–52.

Rankin D.J. and Kokko H. 2007. Do males matter? The role of males in population dynamics. *Oikos*, **116**, 335–48.

Rankin D.J. and López-Sepulcre A. 2005. Can adaptation lead to extinction? *Oikos*, **111**, 616–19.

Ray M. and Stoner A. 1994. Experimental analysis of growth and survivorship in a marine gastropod aggregation: balancing growth with safety in numbers. *Marine Ecology Progress Series*, **105**, 47–59.

Real L.A. 1977. The kinetics of functional response. *American Naturalist*, **111**, 289–300.

Reed D. 2005. Relationship between population size and fitness. *Conservation Biology*, **19**, 563–8.

Reed H.D., O'Grady J.J., Brook B.W., Ballou J.D., and Frankham R. 2003. Estimates of minimum viable population sizes for vertebrates and factors influencing those estimates. *Biological Conservation*, **113**, 23–34.

Reed J. and Dobson A. 1993. Behavioural constraints and conservation biology: conspecific attraction and recruitment. *Trends in Ecology & Evolution*, **8**, 353–5.

Regoes R.R., Ebert D., and Bonhoeffer S. 2002. Dose-dependent infection rates of parasites produce the Allee effect in epidemiology. *Proceedings of the Royal Society of London Series B-Biological Sciences*, **269**, 271–9.

Reluga T. and Viscido S. 2005. Simulated evolution of selfish herd behaviour. *Journal of Theoretical Biology*, **234**, 213–25.

Rex M., McClain C., Johnson N., *et al.* 2005. A source-sink hypothesis for abyssal biodiversity. *American Naturalist*, **165**, 163–78.

Reyes-Bonilla H. and Herrero-Perezrul M. 2003. Population parameters of an exploited population of *Isostichopus fuscus* (Holothuroidea) in the southern Gulf of California, Mexico. *Fisheries Research*, **59**, 423–30.

Rickel S. and Genin A. 2006. Zooplankton aggregations do not confuse planktivorous fish, they can even be beneficial. *EOS, Transactions, American Geophysical Union*, **87**, no. 36, suppl.

Rietkerk M. and Van de Koppel J. 1997. Alternate stable states and threshold effects in semi-arid grazing systems. *Oikos*, **78**, 69–76.

Rietkerk M., Boerlijst M., van Langevelde F., *et al.* 2002. Self-organisation of vegetation in arid ecosystems. *American Naturalist*, **160**, 524–30.

Rietkerk M., Bosch F.V.D., and Van de Koppel J. 1997. Site-specific properties and irreversible vegetation changes in semi-arid grazing systems. *Oikos*, **80**, 241–52.

Rietkerk M., Dekker S., de Ruiter P., and van de Koppel J. 2004. Self-organised patchiness and catastrophic shifts in ecosystems. *Science*, **305**, 1926–9.

Rigler F.H. 1961. The relation between concentration of food and feeding rate of *Daphnia magna* Straus. *Canadian Journal of Zoology*, **39**, 857–68.

Richardson B.J., Hayes R.A., Wheeler S.H., and Yardin M.R. 2002. Social structures, genetic structures and dispersal strategies in Australian rabbit (*Oryctolagus cuniculus*) populations. *Behavioral Ecology and Sociobiology*, **51**, 113–21.

Richman A.D. and Kohn J.R. 1996. Learning from rejection: the evolutionary biology of single-locus incompatibility. *Trends in Ecology & Evolution*, **11**, 497–502.

Ritz D. 2000. Is social aggregation in aquatic crustaceans a strategy to conserve energy? *Canadian Journal of Fisheries and Aquatic Science*, **57** Supplement 3, 59–67.

Rivalan P., Rosser A.M., Delmas V., *et al.* 2007. Can bans stimulate wildlife trade? *Nature*, **447**, 529–30.

Roelke M.E., Martenson J.S., and O'Brien S.J. 1993. The consequences of demographic reduction and genetic depletion in the endangered Florida panther. *Current Biology*, **3**, 340–50.

Roemer G., Coonan T., Garcelon D., Bascompte J., and Laughrin L. 2001. Feral pigs facilitate hyperpredation by golden eagles and indirectly cause the decline of the island fox. *Animal Conservation*, **4**, 307–18.

Roemer G., Donlan C., and Courchamp F. 2002. Golden eagles, feral pigs and insular carnivores: how exotic species turn native predators in to prey. *Proceedings of the National Academy of Sciences of the United States of America*, **99**, 791–6.

Rohlf F.J. 1969. The effect of clumped distributions in sparse populations. *Ecology*, **50**, 716–21.

Rohlfs M. and Hoffmeister T.S. 2003. An evolutionary explanation of the aggregation model of species coexistence. *Proceedings of the Royal Society of London Series B-Biological Sciences*, **270**, S33–S35.

Rohlfs M., Obmann B., and Petersen R. 2005. Competition with filamentous fungi and its implication for a gregarious lifestyle in insects living on ephemeral resources. *Ecological Entomology*, **30**, 556–63.

Rolland C., Danchin E., and de Fraipont M. 1998. The evolution of coloniality in birds in relation to food, habitat, predation and life-history traits: a comparative analysis. *American Naturalist*, **151**, 514–29.

Rose G. 2004. Reconciling overfishing and climate change with stock dynamics of Atlantic cod (*Gadus morhua*) over 500 years. *Canadian Journal of Fisheries and Aquatic Science*, **61**, 1553–7.

Rosenzweig M., Abramsky Z., and Subach A. 1997. Safety in numbers: sophisticated vigilance by Allenby's gerbil. *Proceedings of the National Academy of Sciences of the United States of America*, **94,** 5713–15.

Rosenzweig M.L. 1971. The paradox of enrichment: destabilization of exploitation ecosystems in ecological time. *Science*, **171**, 385–7.

Roughgarden J. and Smith F. 1996. Why fisheries collapse and what to do about it. *Proceedings of the National Academy of Sciences of the United States of America*, **93,** 5078–83.

Rousset F. and Ronce O. 2004. Inclusive fitness for traits affecting metapopulation demography. *Theoretical Population Biology*, **65,** 127–41.

Rowe S., Hutchings J.A., Bekkevold D., and Rakitin A. 2004. Depensation, probability of fertilization, and the mating system of Atlantic cod (*Gadus morhua* L.). *ICES Journal of Marine Science*, **61**, 1144–50.

Russell A.F. 2004. Mammals: comparisons and contrasts. In W.D. Koenig and J.L. Dickinson, eds. *Ecology and evolution of cooperative breeding in birds*, pp. 210–27. Cambridge University Press, Cambridge.

Russell E. and Rowley I. 1988. Helper contributions to reproductive success in the splendid fairy-wren (*Malurus splendens*). *Behavioural Ecology and Sociobiology*, **22,** 131–40.

Ryti R.T. and Case T.J. 1988. The regeneration niche of desert ants: effects of established colonies. *Oecologia*, **75**, 303–6.

Saccheri I., Kuussaari M., Kankare M., Vikman P., Fortelius W., and Hanski I. 1998. Inbreeding and extinction in a butterfly metapopulation. *Nature*, **392**, 491–4.

Sadovy Y. 2001. The threat of fishing to highly fecund fishes. *Journal of Fish Biology*, **59,** 90–108.

Saether B.E., Engen S., and Lande R. 1996. Density-dependence and optimal harvesting of fluctuating populations. *Oikos*, **76,** 40–6.

Saether B.E., Ringsby T.H., and Roskaft E. 1996. Life history variation, population processes and priorities in species conservation: towards a reunion of research paradigms. *Oikos*, **77,** 217–26.

Sakai S. 2002. A review of brood-site pollination mutualism: plants providing breeding sites for their pollinators. *Journal of Plant Research*, **115,** 161–8.

Sakuratani Y., Nakao K., Aoki N., and Sugimoto T. 2001. Effect of population density of *Cylas formicarius* (Fabricius) (Coleoptera: Brentidae) on the progeny populations. *Applied Entomology and Zoology*, **36,** 19–23.

Sanderson E.W. 2006. How many animals do we want to save? The many ways of setting population target levels for conservation. *Bioscience*, **56,** 911–22.

Satake A. and Iwasa Y. 2002. The synchronized and intermittent reproduction of forest trees is mediated by the Moran effect, only in association with pollen coupling. *Journal of Ecology*, **90,** 830–8.

Satake A., Bjoernstad O., and Kobro S. 2004. Masting and trophic cascades: interplay between rowan trees, apple fruit moth, and their parasitoid in southern Norway. *Oikos*, **104,** 540–50.

Satake A., Sasaki A., and Iwasa Y. 2001. Variable timing of reproduction in unpredictable environments: adaption of flood plain plants. *Theoretical Population Biology*, **60,** 1–15.

Savolainen R., Vepsäläinen K., and Deslippe R.J. 1996. Reproductive strategy of the slave ant *Formica podzolica* relative to raiding efficiency of enslaver species. *Insectes Socieaux*, **43**, 201–10.

Seitz R., Lipcius R., Hines A., and Eggleston D. 2001. Density-dependent predation, habitat variation and the persistence of marine bivalve prey. *Ecology*, **82**, 2435–51.

Serrano D. and Tella J. 2007. The role of despotism and heritability in determining settlement patterns in the colonial lesser kestrel. *American Naturalist*, **169**, E53–E67.

Serrano D., Oro D., Esperanza U., and Tella J.L. 2005. Colony size selection determines adult survival and dispersal preferences: Allee effects in a colonial bird. *American Naturalist*, **166**, E22–E31.

Shapiro D. 1989. Sex change as an alternative life-history style. In M. Bruton, ed. *Alternative Life-history Styles of Animals*, pp. 177–95. Kluwer Academic, Dordrecht.

Sharov A., Liebhold A., and Ravlin F. 1995. Prediction of gypsy moth (Lepidoptera: Lymantriidae) mating success from pheromone trap counts. *Environmental Entomology*, **24**, 1239–44.

Shea K. and Possingham H. 2000. Optimal release strategies for biological control agents: an application of stochastic dynamic programming to population management. *Journal of Applied Ecology*, **37**, 77–86.

Shelton P. and Healey B. 1999. Should depensation be dismissed as a possible explanation for the lack of recovery of the northern cod (*Gadus morhua*) stock? *Canadian Journal of Fisheries and Aquatic Science*, **56**, 1521–4.

Shepherd S. and Brown L. 1993. What is an abalone stock? Implications for the role of refugia in conservation. *Canadian Journal of Fisheries and Aquatic Science*, **50**, 2001–9.

Shi J.P. and Shivaji R. 2006. Persistence in reaction diffusion models with weak Allee effect. *Journal of Mathematical Biology*, **52**, 807–29.

Scheffer M. 1990. Multiplicity of stable states in freshwater systems. *Hydrobiologia*, **200**, 475–86.

Scheffer M., Carpenter S., Foley J.A., Folke C., and Walker B. 2001. Catastrophic shifts in ecosystems. *Nature*, **413**, 591–6.

Scheuring I. 1999. Allee effect increases the dynamical stability of populations. *Journal of Theoretical Biology*, **199**, 407–14.

Schlesinger W., Reynolds J., Cunningham G., *et al.* 1990. Biological feedbacks in global desertification. *Science*, **247**, 1043–8.

Schofield P. 2002. Spatially explicit models of Turelli-Hoffmann *Wolbachia* invasive wave fronts. *Journal of Theoretical Biology*, **215**, 121–31.

Schradin C. 2000. Confusion effect in a reptilian and a primate predator. *Ethology*, **106**, 691–700.

Schreiber S.J. 2003. Allee effects, extinctions, and chaotic transients in simple population models. *Theoretical Population Biology*, **64**, 201–9.

Schreiber S.J. 2004. On Allee effects in structured populations. *Proceedings of the American Mathematical Society*, **132**, 3047–53.

Sibly R. 1983. Optimal group size is unstable. *Animal Behaviour*, **31**, 947–8.

Sih A. and Baltus M.S. 1987. Patch size, pollinator behaviour and pollinator limitation in catnip. *Ecology*, **68**, 1679–90.

Sinclair A. and Pech R. 1996. Density dependence, stochasticity, compensation and predator regulation. *Oikos*, **75**, 164–73.

Sinclair A., Pech R., Dickman C., Hik D., Mahon P., and Newsome A. 1998. Predicting effects of predation on conservation of endangered prey. *Conservation Biology*, **12**, 564–75.

Sivashanthini K. and Khan S. 2004. Population dynamics of silver biddy *Gerres setifer* (Pisces: Perciformes) in the Parangipettai waters, southeast coast of India. *Indian Journal of Marine Sciences*, **33**, 346–54.

Soboleva T.K., Shorten P.R., Pleasants A.B., and Rae A.L. 2003. Qualitative theory of the spread of a new gene into a resident population. *Ecological Modelling*, **163**, 33–44.

Solari A., Martin-Gonzalez J., and Bas C. 1997. Stock and recruitment in Baltic cod (*Gadus morhua*): a new, non-linear approach. *ICES Journal of Marine Science*, **54**, 427–43.

Soldaat L.L., Vetter B., and Klotz S. 1997. Sex ratio in populations of *Silene otites* in relation to vegetation cover, population size and fungal infection. *Journal of Vegetation Science*, **8**, 697–702.

Song Y.L., Peng Y.H., and Han M.A. 2004. Travelling wavefronts in the diffusive single species model with Allee effect and distributed delay. *Applied Mathematics and Computation*, **152**, 483–97.

South A.B. and Kenward R.E. 2001. Mate finding, dispersal distances and population growth in invading species: a spatially explicit model. *Oikos*, **95**, 53–8.

Srivastava D. and Jeffries R. 1996. A positive feedback: herbivory, plant growth, salinity and the desertification of an Arctic salt marsh. *Journal of Ecology*, **84**, 31–42.

Steele J.H. and Henderson E.W. 1984. Modelling long-term fluctuations in fish stocks. *Science*, **224**, 985–7.

Steffen C. and Thomas D. 2003. Was Henri Gougerot the first to describe "Hailey-Hailey disease"? *American Journal of Dermatopathology*, **25**, 256–9.

Stephan T. and Wissel C. 1994. Stochastic extinction models discrete in time. *Ecological Modelling*, **75**, 183–92.

Stephens P.A. and Sutherland W.J. 2000. Vertebrate mating systems, Allee effects and conservation. In M. Apollonio, M. Festa-Bianchet and D. Mainardi, eds. *Vertebrate mating systems*, pp. 186–213. World Scientific Publishing, Singapore.

Stephens P.A., Sutherland W.J., and Freckleton R. 1999. What is the Allee effect? *Oikos*, **87**, 185–90.

Stephens P.A. and Sutherland W.J. 1999. Consequences of the Allee effect for behaviour, ecology and conservation. *Trends in Ecology & Evolution*, **14**, 401–5.

Stephens P.A., Frey-Roos F., Arnold W., and Sutherland W.J. (2002a). Model complexity and population predictions. The alpine marmot as a case study. *Journal of Animal Ecology*, **71**, 343–61.

Stephens P.A., Frey-Roos F., Arnold W., and Sutherland W.J. (2002b). Sustainable exploitation of social species: a test and comparison of models. *Journal of Applied Ecology*, **39**, 629–42.

Stephens P.A., Buskirk S.W., and Martinez del Rio C. 2007. Inference in ecology and evolution. *Trends in Ecology & Evolution*, **22**, 192–7.

Stevens E. and Pickett C. 1994. Managing the social environments of flamingoes for reproductive success. *Zoo Biology*, **13**, 501–7.

Stevens M.A. and Boness D.J. 2003. Influences of habitat features and human disturbance on use of breeding sites by a declining population of southern fur seals (*Arctocephalus australis*). *Journal of Zoology*, **260**, 145–52.

Stewart-Cox J.A., Britton N.F. and Mogie M. 2005. Pollen limitation or mate search need not induce an Allee effect. *Bulletin of Mathematical Biology*, **67**, 1049–79.

Stigter J.D. and van Langevelde F. 2004. Optimal harvesting in a two-species model under critical depensation—The case of optimal harvesting in semi-arid grazing systems. *Ecological Modelling*, **179**, 153–61.

Stiling P.D. 1987. The frequency of density dependence in insect host-parasitoid systems. *Ecology*, **68**, 844–56.

Stoddart D.M. 1974. Saiga find safety in numbers. *Nature*, **250**, 11–12.

Stoner A. and Ray M. 1993. Aggregation dynamics in juvenile queen conch (*Strombus gigas*): population structure, mortality, growth and migration. *Marine Biology*, **116**, 571–82.

Stoner A.W. and Ray-Culp M. 2000. Evidence for Allee effects in an over-harvested marine gastropod: density-dependent mating and egg production. *Marine Ecology-Progress Series*, **202**, 297–302.

Stuart B., Rhodin A., Grismer L., and Hansel T. 2006. Scientific description can imperil species. *Science*, **312**, 1137.

Studer-Thiersch A. 2000. What 19 years of observation on captive greater flamingoes suggests about adaptations to breeding under irregular conditions. *Waterbirds*, **23**, 150–9.

Sugg D.W. and Chesser R.K. 1994. Effective population sizes with multiple paternity. *Genetics*, **137**, 1147–55.

Sugie J. and Kimoto K. 2006. Homoclinic orbits in predator–prey systems with a nonsmooth prey growth rate. *Quarterly of Applied Mathematics*, **64**, 447–61.

Sun Y.G. and Saker S.H. 2005a. Existence of positive periodic solutions of nonlinear discrete model exhibiting the Allee effect. *Applied Mathematics and Computation*, **168**, 1086–97.

Sun Y.G. and Saker S.H. 2005b. Oscillatory and asymptotic behavior of positive periodic solutions of nonlinear discrete model exhibiting the Allee effect. *Applied Mathematics and Computation*, **168**, 1205–18.

Sutherland W.J. 2002. Conservation biology—science, sex and the kakapo. *Nature*, **419**, 265–6.

Sutherland W.J. 2003. Parallel extinction risk and global distribution of languages and species. *Nature*, **423**, 276–9.

Swain D. and Sinclair A. 2000. Pelagic fishes and the cod recruitment dilemma in the Northwest Atlantic. *Canadian Journal of Fisheries and Aquatic Science*, **57**, 1321–5.

Swart J., Lawes M.J. and Perrin M.R. 1993. A mathematical model to investigate the demographic viability of low-density samango monkey (*Cercopithecus mitis*) populations in Natal, South Africa. *Ecological Modelling*, **70**, 289–303.

Takeuchi Y. 1996. *Global dynamical properties of Lotka-Volterra systems*. World Scientific Publishing Company, Singapore.

Taylor A.D. 1990. Metapopulations, dispersal and predator–prey dynamics: an overview. *Ecology*, **71**, 429–33.

Taylor C.M. and Hastings A. 2004. Finding optimal control strategies for invasive species: a density-structured model for *Spartina alterniflora*. *Journal of Applied Ecology*, **41**, 1049–57.

Taylor C.M. and Hastings A. 2005. Allee effects in biological invasions. *Ecology Letters*, **8**, 895–908.

Taylor C.M., Davis H.G., Civille J.C., Grevstad F.S., and Hastings A. 2004. Consequences of an Allee effect in the invasion of a pacific estuary by *Spartina alterniflora*. *Ecology*, **85**, 3254–66.

Tcheslavskaia K., Brewster C.C., and Sharov A.A. 2002. Mating success of gypsy moth (Lepidoptera: Lymantriidae) females in Southern Wisconsin. *Great Lakes Entomologist*, **35**, 1–7.

Tegner M. and Dayton P. 1977. Sea urchin recruitment patterns and implications of commercial fishing. *Science*, **196**, 324–6.

Tegner M., Basch L., and Dayton P. 1996. Near extinction of a marine invertebrate. *Trends in Ecology & Evolution*, **11**, 278–80.

Tegner M., DeMartini J., and Karpov K. 1992. The California red abalone fishery: a case study in complexity. In S. Shepherd, M. Tegner and G.D. Proo, eds. *Abalone of the world: biology, fisheries and culture*, pp. 370–83. Fishing News Books, Oxford.

Thieme H.R. 2003. *Mathematics in population biology*. Princeton University Press, Princeton and Oxford.

Thomas J. and Benjamin M. 1973. The effects of population density on growth and reproduction of *Biophalaria glabrata* (Say) (Gastropoda: Pulmonata). *Journal of Animal Ecology*, **43**, 31–50.

Thompson D., Strange I., Riddy M., and Duck C. 2005. The size and status of the population of southern sea lions *Otaria flavescens* in the Falkland Islands. *Biological Conservation*, **121**, 357–67.

Tobin P.C., Whitmire S.L., Johnson D.M., Bjørnstad O.N., and Liebhold A.M. 2007. Invasion speed is affected by geographical variation in the strength of Allee effects. *Ecology Letters*, **10**, 36–43.

Tosh C., Jackson A., and Ruxton G. 2006. The confusion effect in predatory neural networks. *American Naturalist*, **167**, E52–E58.

Travis J.M.J. and Dytham C. 2002. Dispersal evolution during invasions. *Evolutionary Ecology Research*, **4**, 1119–29.

Treves A. 2000. Theory and method in the study of vigilance and aggregation. *Animal Behaviour*, **60**, 711–22.

Treyde A., Williams J., Bedin E., *et al.* 2001. In search of the optimal management strategy for Arabian oryx. *Animal Conservation*, **4**, 239–49.

Trinkel M. and Kastberger G. 2005. Competitive interactions between spotted hyenas and lions in the Etosha National Park, Namibia. *African Journal of Ecology*, **43**, 220–4.

Troell M., Robertson-Andersson D., Anderson R., *et al.* 2006. Abalone farming in South Africa: an overview with perspectives on kelp resources, abalone feed, potential for on-farm seaweed production and socio-economic importance. *Aquaculture*, **257**, 266–81.

Tschinkel W.R. 1993. Sociometry and sociogenesis of colonies of the fire ant Solenopsis invicta during one annual cycle. *Ecological Monographs*, **63**, 425–57.

Turelli M. and Hoffmann A.A. 1991. Rapid spread of an inherited incompatibility factor in California *Drosophila*. *Nature*, **353**, 440–2.

Turelli M. and Hoffmann A.A. 1995. Cytoplasmic incompatibility in *Drosophila simulans*: dynamics and parameter estimates from natural populations. *Genetics*, **140**, 1319–38.

Turesson H. and Broenmark C. 2004. Foraging behaviour and capture success in perch, pike-perch and pike and the effects of prey density. *Journal of Fish Biology*, **65**, 363–75.

Turchin P. 2003. *Complex population dynamics. A theoretical/empirical synthesis*. Princeton University Press, Princeton, New Jersey.

Turchin P. and Kareiva P. 1989. Aggregation in *Aphis varians*: an effective strategy for reducing predation risk. *Ecology*, **70**, 1008–16.

Tzeng T.-D., Chiu C.-S., and Yeh S.-Y. 2005. Growth and mortality of the red-spot prawn (*Metapenaeopsis barbata*) in the northeastern coast off Taiwan. *Journal of the Fisheries Society of Taiwan*, **32**, 229–38.

van de Koppel J., Herman P.M.J., Thoolen P., and Heip C.H.R. 2001. Do alternate stable states occur in natural ecosystems? Evidence from a tidal flat. *Ecology*, **82**, 3449–61.

van de Koppel J., Rietkerk M., Dankers N., and Herman P.M.J. 2005. Scale-dependent feedback and self-organized patterns in young mussel beds. *American Naturalist*, **615**, E66–E77.

van Kooten T., de Roos A.M., and Persson M. 2005. Bistability and an Allee effect as emergent consequences of stage-specific predation. *Journal of Theoretical Biology*, **237**, 67–74.

Van Rossum F., Campos de Sousa S. and Triest L. 2004. Genetic consequences of habitat fragmentation in an agricultural landscape on the common *Primula veris*, and comparison with its rare congener, *P. vulgaris*. *Conservation Genetics*, **5**, 231–45.

Van Valen L. 1973 A New Evolutionary Law. *Evolutionary Theory*, **1**, 1–30.

Vargas S.F. 2006. Breeding behavior and colonization success of the Cuban treefrog *Osteopilus septentrionalis*. *Herpetologica*, **62**, 398–408.

Veit R.R. and Lewis M.A. 1996. Dispersal, population growth, and the Allee effect: dynamics of the house finch invasion of eastern North America. *American Naturalist*, **148**, 255–74.

Veltman C., Nee S., and Crawley M. 1996. Correlates of introduction success in exotic New Zealand birds. *American Naturalist*, **147**, 542–57.

Vernon J. 1995. Low reproductive output of isolated, self-fertilising snails: inbreeding depression or absence of social facilitation? *Proceedings of the Royal Society of London B*, **259**, 131–6.

Vogel H., Czihak G., Chang P., and Wolf W. 1982. Fertilization kinetics of sea urchin eggs. *Mathematical Biosciences*, **58**, 189–216.

Volkov I., Banavar J.R., He F., Hubbell S.P., and Maritan A. 2005. Density dependence explains tree species abundance and diversity in tropical forests. *Nature*, **438**, 658–61.

Volterra V. 1938. Population growth, equilibria, and extinction under specified breeding conditions: a development and extension of the logistic curve. *Human Biology*, **3**, 3–11.

Vucetich J., Peterson R., and Waite T. 2004. Raven scavenging favours group foraging in wolves. *Animal Behaviour*, **67**, 1117–26.

Wagenius S. 2006. Scale dependence of reproductive failure in fragmented *Echinacea* populations. *Ecology*, **87**, 931–41.

Waldman J.R., Bender R.E., and Wirgin I.I. 1998. Multiple population bottlenecks and DNA diversity in populations of wild striped bass, *Morone saxatilis*. *Fishery Bulletin*, **96**, 614–20.

Walters C. and Kitchell J.F. 2001. Cultivation/depensation effects on juvenile survival and recruitment: implications for the theory of fishing. *Canadian Journal of Fisheries and Aquatic Sciences*, **58**, 39–50.

Walters J.R., Crowder L.B., and Priddy J.A. 2002. Population viability analysis for red-cockaded woodpeckers using an individual-based model. *Ecological Applications*, **12**, 249–60.

Wang G., Liang X.G., and Wang F.Z. 1999. The competitive dynamics of populations subject to an Allee effect. *Ecological Modelling*, **124**, 183–92.

Wang M.H. and Kot M. 2001. Speeds of invasion in a model with strong or weak Allee effects. *Mathematical Biosciences*, **171**, 83–97.

Wang M.H., Kot M., and Neubert M.G. 2002. Integrodifference equations, Allee effects, and invasions. *Journal of Mathematical Biology*, **44**, 150–68.

Warner R. 1984. Mating behaviour and hermaphroditism in coral reef fishes. *American Scientist*, **72**, 128–36.

Waser P., Elliott L., Creel N., and Creel S. 1995. Habitat variation and mongoose demography. In A. Sinclair and P. Arcese, eds. *Serengeti II: dynamics, management and conservation of an ecosystem*, pp. 421–48. University of Chicago Press, Chicago.

Waters D., Noble R., and Hightower J. 2005. Fishing and natural mortality of adult largemouth bass in a tropical reservoir. *Transactions of the American Fisheries Society*, **134**, 563–71.

Webb C. 2003. A complete classification of Darwinian extinction in ecological interactions. *American Naturalist*, **161**, 181–205.

Welch A.M., Semlitsch R.D., and Gerhardt H.C. 1998. Call duration as an indicator of genetic quality in male gray tree frogs. *Science*, **280**, 1928–30.

Wells H., Strauss E.G., Rutter A.M., and Wells P.H. 1998. Mate location, population growth and species extinction. *Biological Conservation*, **86**, 317–24.

Wells H., Wells P.H., and Cook P. 1990. The importance of overwinter aggregation for reproductive success of monarch butterflies (*Danaus prexippus* L.). *Journal of Theoretical Biology*, **147**, 115–31.

Wertheim B., Marchais J., Vet L.E.M., and Dicke M. 2002. Allee effect in larval resource exploitation in *Drosophila*: an interaction among density of adults, larvae, and microorganisms. *Ecological Entomology*, **27**, 608–17.

Weyl O., Booth A., Mwakiyongo K., and Mandere D. 2005. Management recommendations for *Copadichromis chrysonotus* (Pisces: Cichlidae) in Lake Malombe, Malawi, based on per-recruit analysis. *Fisheries Research*, **71**, 165–73.

Whitmire S. and Tobin P. 2006. Persistence of invading gypsy moth populations in the United States. *Oecologia*, **147**, 230–7.

Widén B. 1993. Demographic and genetic effects on reproduction as related to population size in a rare, perennial herb, *Senecio integrifolius* (Asteraceae). *Biological Journal of the Linnean Society*, **50**, 179–95.

Wiegand K., Henle K., and Sarre S.D. 2002. Extinction and spatial structure in simulation models. *Conservation Biology*, **16**, 117–28.

Wiernasz D.C. and Cole B.J. 1995. Spatial distribution of *Pogonomyrmex occidentalis*: recruitment, mortality and overdispersion. *Journal of Animal Ecology*, **64**, 519–27.

Wiklund C. and Fagerström T. 1977. Why do males emerge before females—hypothesis to explain incidence of protandry in butterflies. *Oecologia*, **31**, 153–8.

Wilcock C. and Neiland R. 2002. Pollination failure in plants: why it happens and when it matters. *Trends in Plant Science*, **7**, 270–7.

Wilhelm F.M., Schindler D.W., and McNaught A.S. 2000. The influence of experimental scale on estimating the predation rate of *Gammarus lacustris* (Crustacea: Amphipoda) on *Daphnia* in an alpine lake. *Journal of Plankton Research*, **22**, 1719–34.

Willi Y., Van Buskirk J., and Fischer M. 2005. A threefold genetic Allee effect: population size affects cross-compatibility, inbreeding depression and drift load in the self-incompatible *Ranunculus reptans*. *Genetics*, **169**, 2255–65.

Wilson E.O. and Bossert W.H. 1971. *A primer of population biology*. Sinauer Associates, Sunderland, Massachusetts.

Wilson J. and Agnew A. 1992. Positive feedback switches in plant communities. *Advances in Ecological Research*, **23**, 263–336.

Wilson W.G. and Harder L.D. 2003. Reproductive uncertainty and the relative competitiveness of simultaneous hermaphroditism versus dioecy. *American Naturalist*, **162**, 220–41.

Wittmer H., Sinclair A., and McLellan B. 2005. The role of predation in the decline and extirpation of woodland caribou. *Oecologia*, **144**, 257–67.

Wolf C., Garland T., and Griffith B. 1998. Predictors of avian and mammalian translocation success: reanalysis with phylogenetically independent contrasts. *Biological Conservation*, **86**, 243–55.

Wolf C., Griffith B., Reed C., and Temple S. 1996. Avian and mammalian translocations: update and reanalysis of 1987 survey data. *Conservation Biology*, **10**, 1142–54.

Wood C. 1987. Predation of juvenile Pacific salmon by the common merganser (*Mergus merganser*) on eastern Vancouver Island. I. Predation during the seaward migration. *Canadian Journal of Fisheries and Aquatic Science*, **44**, 941–9.

Wootton J. 1994. The nature and consequences of indirect effects in ecological communities. *Annual Review of Ecology and Systematics*, **25**, 443–66.

Wray G.A. 1995. Evolution of larvae and developmental modes. In L.R. McEdwards, ed. *Ecology of marine invertebrate larvae*, pp. 413–47. CRC Press, Boca Raton, Florida.

Wright J. 1988. Helpers-at-the-nest have the same provisioning rule as parents: experimental evidence from playbacks of chick begging. *Behavioural Ecology and Sociobiology*, **42**, 423–9.

Xiao Y. and McShane P. 2000. Use of age- and time-dependent seasonal growth models in analysis of tag/recapture data on the western king prawn *Penaeus latsulcatus* in the Gulf of St. Vincent, Australia. *Fisheries Research*, **49**, 85–92.

Yan J.R., Zhao A.M., and Yan W.P. 2005. Existence and global attractivity of periodic solution for an impulsive delay differential equation with Allee effect. *Journal of Mathematical Analysis and Applications*, **309**, 489–504.

Yoshimura A., Kawasaki K., Takasu F., Togashi K., Futai K., and Shigesada N. 1999. Modeling the spread of pine wilt disease caused by nematodes with pine sawyers as vector. *Ecology*, **80**, 1691–1702.

Young A., Boyle T., and Brown T. 1996. The population genetic consequences of habitat fragmentation for plants. *Trends in Ecology & Evolution*, **11**, 413–18.

Young A.G., Brown A.H.D., Murray B.G., Thrall P.H., and Miller C.H. 2000. Genetic erosion, restricted mating and reduced viability in fragmented populations of the endangered grassland herb Rutidosis leptorrhynchoides. In A.G. Young and G.M. Clarke, eds. *Genetics, demography and viability of fragmented populations*, pp. 335–59. Cambridge University Press, Cambridge.

Zhou S.R. and Wang G. 2004. Allee-like effects in metapopulation dynamics. *Mathematical Biosciences*, **189**, 103–13.

Zhou S.R. and Wang G. 2006. One large, several medium, or many small? *Ecological Modelling*, **191**, 513–20.

Zhou S.R. and Zhang D.Y. 2006. Allee effects and the neutral theory of biodiversity. *Functional Ecology*, **20**, 509–13.

Zhou S.R., Liu C.Z., and Wang G. 2004. The competitive dynamics of metapopulations subject to the Allee-like effect. *Theoretical Population Biology*, **65**, 29–37.

Zhou S.R., Liu Y.F., and Wang G. 2005. The stability of predator–prey systems subject to the Allee effects. *Theoretical Population Biology*, **67**, 23–31.

Zhu J. and Qiu Y. 2005. Growth and mortality of hairtails and their dynamic pool models in the northern South China Sea. *Acta Oceanologica Sinica*, **27**, 93–9.

Zmora O., Findiesen A., Stubblefield J., Frenkel V., and Zohar Y. 2005. Large-scale juvenile production of the blue crab *Callinectes sapidus*. *Aquaculture*, **244**, 129–39.

Index